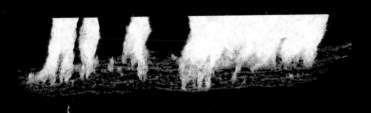

collen

200 Years of Building and
Civil Engineering in Ireland

collen

200 Years of Building and Civil Engineering in Ireland

A History of the Collen Family Business
1810–2010

JOHN WALSH

THE LILLIPUT PRESS
DUBLIN

First published 2010 by
THE LILLIPUT PRESS
62–63 Sitric Road, Arbour Hill
Dublin 7, Ireland
www.lilliputpress.ie

Copyright © John Walsh, 2010

ISBN 978 1 84351 176 2

1 3 5 7 9 10 8 6 4 2

A CIP record for this title is available
from The British Library.

Set in 11 pt on 14.5 pt Garamond with Gotham titling by Marsha Swan
Printed and bound in the UK by J.F. Print Ltd, Sparkford, Somerset

Contents

Acknowledgments

This book would not have happened without the interest and support of a considerable range of people. I owe a debt of gratitude to Neil Collen for initiating the project and to Prof. David Dickson for his advice and encouragement. I wish to thank Prof. Eunan O'Halpin of the Centre of Contemporary Irish History, Trinity College, which provided my official home while completing the project. Thanks are also due to Ellen Hanley of the Trinity Foundation and to Pamela Hilliard and Judith Lee in the History Department.

Archivists and librarians at various institutions did a great deal to facilitate my research. I wish to thank the staff of Trinity College Dublin Library, Blanchardstown Library, the National Library of Ireland and the Companies Registration Office. I am particularly grateful to Caitriona Crowe at the National Archives and all the staff of the Irish Architectural Archive for their invaluable advice and assistance.

Members of the Collen family gave generously of their time and recollections. I wish to thank Neil and David Collen for discussing the recent history of the company in Dublin; Niall and Joe Collen for their insights on the fortunes of the firm in Portadown. Peter Collen unearthed a wide range of papers and images on Collen Brothers Quarries and previous incarnations of the firm. The assistance of directors and employees of the company, past and present, was also crucial to the timely completion of the project. Dr. Brian Bond was consistently helpful and informative. Kara Craig and Sinéad Savage proved adept at finding old company accounts and records. I am grateful to Sandra Muller for preparing the illustrations for publication. I owe a particular debt of gratitude to Paddy Wall, who provided invaluable advice and dealt assiduously with a multitude of queries about the history of the company.

I wish to express my appreciation to all those who agreed to be interviewed for this book, including Dr Ronald Cox, Dr Brian Bond, Marcus Collie, Seamus Small, Martin Glynn, Leo Crehan, Des Lynch, John Ruane, Louise Coffey, Rita McMillan, Chris Lyons, Pat Sides, Frank O'Sullivan, and Jerry O'Leary.

I wish to thank Antony Farrell and Lilliput Press for publishing the book. Marsha Swan, Djinn Van Noorden and Kitty Lyddon all did a great deal to prepare the work for publication. Lisa Scholey prepared the index to her usual high standard. I am particularly grateful to Fiona Dunne, who offered invaluable encouragement and expertise. I am greatly indebted to Lorna Moloney for preparing the family tree.

Finally friends and family members who helped and advised throughout the lengthy process of writing and rewriting deserve my profound gratitude. Thanks are due to Siobhan and Kevin for their tolerance of constant queries and skilled technical assistance. I wish to record my thanks especially to my parents, Maura and John Walsh, whose consistent help and encouragement has been essential to the successful completion of this work.

John Walsh
August 2010

Introduction

This book sets out to trace the evolution of the Collen family business, from its origins in early nineteenth-century Ulster, through the division of the firm between north and south, to the development of the modern companies operating in Dublin and Portadown. The main objective of the work is to explore the development of the firm in changing political, economic and social contexts. The influence of the distinctive values fostered by the family proprietors is a consistent theme, not least because the study suggests that the nature of the firm as a family run institution contributed to its stability, longevity and commercial success. The book evaluates how a family run enterprise adapted to the far-reaching transformation of politics and society in Ireland over a period of two centuries.

The study is based principally on archival material not previously available or exploited for research purposes. The collection of business records held by Collen Construction at River House in East Wall Road offered a wide variety of commercial information from 1880 to the present day, including correspondence, ledgers and two sales books detailing the firm's contracts over several decades. I was fortunate to discover a rich collection of papers in the offices of Collen Brothers Quarries in Portadown, which was the main repository of the firm's records up to 1949; it was a treasure trove in particular for the era preceding the division of the company. The work also makes use of private family papers belonging to members of the Collen family in both Dublin and Portadown. The papers of Lyal Collen are the most extensive of these private collections, consisting of a range of letters, handwritten notes and photographs, as well as some preliminary comments on the history of the company. The private papers of Joe Collen, including notes and an address to the Rotary Society in Portadown in 1971, are an equally valuable and important source. The records of architectural practices and quantity surveyors, preserved by the Irish Architectural Archives, provide a wealth of fascinating detail on public and private building projects in nineteenth- and early twentieth-century Ireland. The book draws heavily on the back issues of specialized trade journals, especially the *Irish Builder* and *The Architect's Journal*; it also makes use of national newspapers and publications of local history societies. State records were consulted extensively, including the 1911 census in the National Archives and various records in the Companies Registration Office. I also conducted a considerable range of interviews with family members and current or former employees of the two companies, which were used in conjunction with the available documentary sources.

The concept of family business has attracted considerable scholarly interest from economists and economic historians. Yet Andrea Colli, in his study of family business in a historical and comparative perspective, notes that recent research on family firms has become multi-disciplinary, drawing upon politics, sociology and management as much as on history and economics.[1] Colli acknowledges that a specific definition of family business is elusive, but suggests a valuable descriptive model for a family enterprise: the 'classic' family firm is one in which ownership and control are firmly intertwined, where family members are deeply involved in both strategic and day-to-day decision-making and the business is shaped by a 'dynastic' or familial motive.[2] Much of the academic analysis of the family firm during the 1960s and 1970s was undertaken in the context of wider studies on the development of managerial capitalism and treated family firms as an initial stage on the way to the development of more advanced organizational structures. Alfred Chandler traced the shift from personal or family capitalism to managerial capitalism in developed economies from the middle of the nineteenth century, marked

---

1. Andrea Colli, *The History of Family Business: New Studies in Economic and Social History* (Cambridge, 2003), p.26.

2. *Ibid.* p.9.

in particular by the rise of the managerial corporation. Chandler identified a gradual separation between ownership and control in major corporations, in which top-level decision-making was taken over by professional managers, so that family shareholders ceased to control the business and retained influence only over the income which they derived from it.[3] The managerial corporation was characterized by a wide range of operating units undertaking a complex range of activities and the emergence of increasingly sophisticated hierarchies of salaried managers.[4] The development of managerial enterprise, in particular the emergence of the public company, exerted a profound influence on the evolution of global capitalism during the twentieth century. The transition from family business to the managerial corporation was often presented by scholars as an inevitable process of change driven by financial and technological pressures. The apparent predominance of managerial capitalism generated an extensive literature, produced by academics and consultants, which tended to portray family firms as relatively backward, lacking in dynamism and less competitive than the more advanced managerial corporations.[5]

Yet a very different academic perspective on family business has emerged more recently. The process of evolution from family business to managerial capitalism undoubtedly occurred, particularly in the USA and Britain, but it was not a universal phenomenon. Recent studies have recognized the persistence of family business in modern advanced economies, both in Western Europe, where it may be associated with a distinctive continental model of capitalism that reveals a considerable divergence with the Anglo-Saxon model, and in East Asia, where successful family run conglomerates played a central part in a relatively late process of industrialization.[6] Moreover, new research has highlighted the contribution of family firms to economic development and considered the influence of the particular institutional and historical environment in facilitating the persistence of family business.[7] Indeed, Colli argues that the lasting presence of a familial form of business organization may well

---

3. Alfred Chandler, 'The United States: Seedbed of Managerial Capitalism', in A. Chandler and H. Daems (ed), *Managerial Hierarchies: Comparative Perspectives on the Rise of the Modern Industrial Enterprise* (Harvard, 1980), pp.13–14.

4. *Ibid.*

5. Colli, *History of Family Business*, pp.21–3.

6. Colli, *History of Family Business*, p.26; *Ibid.* p.31.

7. Andrea Colli and Mary Rose, *Families and Firms: The Culture and Evolution of Family Firms in Britain and Italy in the Nineteenth and Twentieth Centuries*, *Scandinavian Economic History Review*, vol.47. no.1 (1999), pp.24–47.

be regarded as the 'the best demonstration of its "efficiency" against a defined institutional framework…'[8] While academic debate and investigation in the area are still continuing, it is apparent that the family firm is an enduring element in modern industrial capitalism rather than simply a transitory stage on the way to a more impersonal or sophisticated model of organization.

The present study considers the history of the Collen family enterprise against the background of the institutional environment in which it operated, exploring the political, economic and social realities that shaped the context for the development of the business. From the outset the firm displayed the key features of a 'classic' family business, notably the close identification of the interests of the family with the business, direct control of the enterprise by family members, a distinctive pattern of internal succession and a dependence on local production factors.[9] Collen would retain the essential characteristics of a family business through two World Wars, the partition of the island, a world economic depression and the Troubles. The division of the original firm in 1949, which saw the creation of two independent companies based in Portadown and Dublin, did not see any dilution in the central position of the Collen family on either side of the border. Yet if continuity of ownership and control marked a defining feature of the family enterprise, the proprietors also embraced the benefits of technical innovation, technological change and managerial expertise. The shareholders showed considerable flexibility and pragmatism in responding to external pressures and sustaining the commercial strength of their business. They did not hesitate to undertake significant organizational change, including the division of the business itself but also measures for rationalization and far-reaching structural reform at various times. The Collen family enterprise showed a striking resilience in adapting to radical changes in its political and economic environment, while maintaining a high level of stability in its ownership and managerial structures.

---

8. Colli, *History of Family Business*, p.26.

9. *Ibid.* p.9.

Collen Family Tree 1782-2009

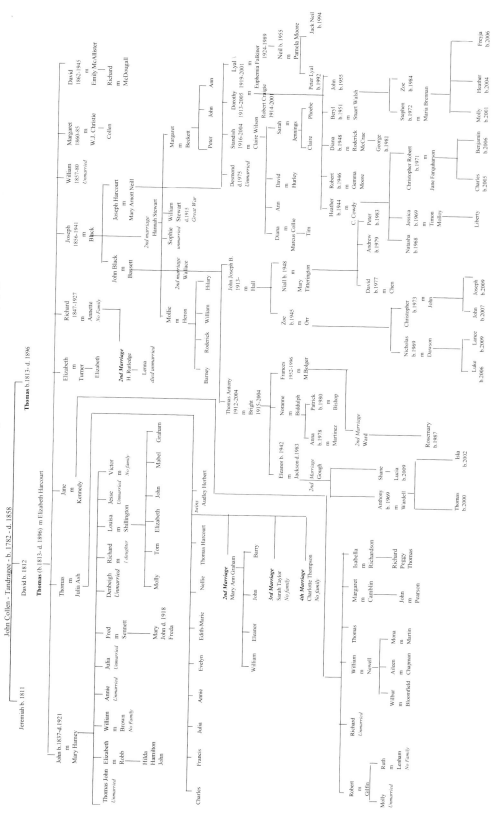

collen

200 Years of Building and
Civil Engineering in Ireland

ONE

Founding the Business

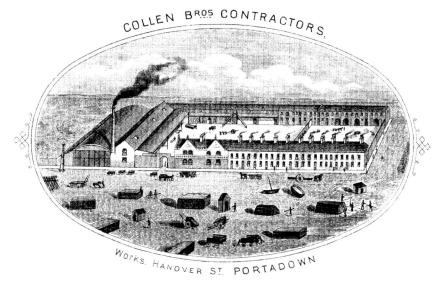

Original builders' yard and offices of Collen Brothers, Hanover St, Portadown, c.1880.

The first record of involvement by the Collen family in the building trade emerges with John Collen. Born in 1782, John Collen was a stonemason who lived and worked in the townland of Knockmore, in the Barony of Ballinamore, Co. Armagh.[1] The family soon became associated with the nearby town of Tandragee. The details of John Collen's life are largely unknown, but he was regarded by subsequent generations of the Collen family as the original founder of the business. Lyal Collen, writing in the 1990s, described him as 'the progenitor of a building family which has survived the vicissitudes of

---

1. Lyal Collen Papers, *The Collen Building Tradition*, p.1.

two centuries…'[2] This is essentially a fair assessment, although the foundation of Collen Brothers as an institutional entity occurred considerably later and owed a great deal to the efforts of John Collen's grandson and namesake. The family business emerged in one of the most industrialized areas of the country: north Armagh was the centre of the linen economy in Ireland during the nineteenth century. Tandragee was a prosperous market town, boasting a population of 1200 as early as 1820.[3] The weekly market was one of the largest in the county and was particularly well known for the quality of its linen products.[4] Thomas Bradshaw commented favourably on Tandragee in his *General Directory* of the towns of Down and Armagh, published in 1820, noting that: 'The town contains a great number of well supplied shops and has been, for some years, increasing in business and respectability.'[5] The density of the population was relatively high and there was a considerable demand for modest building projects. It was no accident that the origins of a flourishing building enterprise were to be found in north Armagh.

The elder John Collen was active in the building trade from the first decade of the nineteenth century: his only son Thomas, who was born in 1813, followed in his father's footsteps. His earliest known contract was the construction of a bridge over the Newry Canal in 1836.[6] Thomas undertook various building projects in Armagh and also managed a small quarrying business.[7] While Thomas was himself one of only two children, he and his wife Eliza Harcourt raised a family of nine; they lived at Ballyknock between Tandragee and the village of Laurelvale. All of the male children were trained tradesmen with the exception of their youngest son David and all became involved in the building business.[8] It appears that John Collen and his son were essentially independent tradesmen who laid the foundations for a fledgling building enterprise during the early nineteenth century. They were not exclusively involved in construction and it is unlikely that they regarded themselves solely as builders. Yet the tradition of Collen involvement in the building trade was firmly established

2. *Ibid.*

3. Thomas Bradshaw, *Bradshaw's General Directory of Newry, Armagh and the towns adjoining for 1820* (Newry, 1820), pp.95–7.

4. *Ibid.* p.97.

5. *Ibid.*

6. Collen Papers Portadown, Speech by Joe Collen to the Rotary Society, *My Job*, March 1971, p.1.

7. Lyal Collen Papers, *Collen: An Irish Building Family 1782–1992*, p.1.

8. Lyal Collen Papers, *Note on Collen family*, 1970.

Thomas and Eliza Collen, Tandragee, 1894–95.
Courtesy of Collen family.

in the first half of the nineteenth century, although the family's activity was restricted to Tandragee and the surrounding area.

A distinctive Collen construction business emerged in the following generation, directed in the first instance by Thomas' eldest son John. The younger John Collen (1837–1921) was the driving force behind the creation of Collen Brothers. John, who showed great ambition and commercial flair from the outset of his career, emigrated to Australia in 1858 shortly after the completion of his apprenticeship.[9] John and his younger brother Thomas undertook the arduous four-month journey to the southern hemisphere in search of work in the building trade. The brothers settled in Melbourne, where they secured employment without much difficulty.[10] John, who was the more dynamic of the two, soon established himself as a building contractor. He secured considerable expertise

9. *Collen: An Irish Building Family 1782–1992*, p.1.

10. Speech by Joe Collen to the Rotary Society, *My Job*, March 1971, p.1.

in the construction business during his time in Australia, and perhaps more significantly, he acquired the capital that made possible his return to Ireland as a prosperous entrepreneur. The young Irishman secured a contract to build a warehouse in Melbourne, but soon became embroiled in a dispute with the local architect on the project, who alleged that the foundations provided by the contractor were flawed and demanded that the work be undertaken again from the beginning.[11] John Collen refused to comply, and family tradition provides a colourful and somewhat rose-tinted description of what happened next. The account given by his grandnephew Lyal suggests that John consulted the local Methodist minister and left the site in accordance with his advice. He then discovered, fortuitously, that the construction of the warehouse was proceeding smoothly on the same foundations that he had first provided. The young contractor took a legal case against the architect and secured a financial settlement compensating him for loss of earnings and the building work that he had already completed.[12] This account was designed to emphasize John's integrity, piety and good character. There is no reason to doubt that such a dispute took place, but it is likely that the rough edges of the story have been smoothed over. Other notes in the Collen papers suggest that John Collen was confronted with an unpalatable choice between rebuilding the foundations at his own expense or dismissal from the job. He walked off the job rather than meet the onerous demands of the architect and perhaps the client; it is evident that he was effectively forced to abandon the contract, due to ineptitude and perhaps hostility on the part of the architect. While he was undoubtedly a committed Methodist, John's actions were dictated not by the advice of his minister but by clear-headed business acumen and force of character. He took the opportunity to turn the tables on the architect, securing a substantial award of compensation in the Australian courts. The dispute, which might have proved damaging to his professional career in different circumstances and could easily have demoralized a less forceful individual, turned out favourably for the young entrepreneur.

The successful outcome of the legal case paved the way for John Collen's return to Ireland. Shortly after the dispute was resolved in his favour, Collen decided to return to his native Armagh, with the intention of establishing his own construction business; his motivation for a rapid return to Ireland was undoubtedly influenced by the increased capital at his disposal following the court proceedings. He saw a golden opportunity to transform the family's

---

11. *Collen: An Irish Building Family 1782–1992*, p.2.

12. *Ibid.* p.2.

Advertisement by Thomas Collen, Bloomhill, Tandragee: 'Hauling, Threshing, Stone-breaking, Hay-Pressing done at shortest notice'. Courtesy of Niall Collen.

long-standing involvement in the building trade into a cohesive business enterprise, making full use of the existing system of canals and the emerging network of railways in Ulster.[13] John Collen took the lead in establishing a new family company, drawing in his brother Richard and their younger siblings Joseph and David. Collen Brothers opened their doors for business as building contractors in Carlton Street, Portadown, in 1867.[14]

Thomas Collen, who was John's compatriot in Australia and returned to Ireland at the same time, did not become a founding partner of the new firm. Instead he rejoined his father in Tandragee and later took charge of the original building business established during the previous two generations.[15] Thomas Collen & Son was active in the building trade in east Ulster until the early twentieth century; one of its most notable projects was the rebuilding of a Methodist church in Armagh town in 1888, while the younger man also oversaw local drainage schemes and the building of labourers' cottages early in the 1900s.[16] Thomas maintained his involvement in the building trade following his father's death in 1896, but it was a relatively small-scale business with an unmistakable regional focus; his activity was largely restricted to Armagh and east Tyrone. Thomas, like his father and grandfather, was

---

13. *Ibid.*

14. Speech by Joe Collen to the Rotary Society, *My Job*, March 1971, p.1.

15. *Ibid.*; Note by Joe Collen, *Collen Bros.*, p.1; *Irish Times*, 'Marriages', 24 August 1888.

16. *The Irish Builder and Engineering Record*, vol.30, 'Armagh Methodist Church', 1 December 1888, p.301.

Court House, Portadown. Courtesy of Niall Collen.

not solely a building contractor: he was also a local landowner and farmer in Tandragee, who became a Justice of the Peace (JP) towards the end of his career.[17] It was the new firm founded in 1867 which would establish the Collen family enterprise on a wider national stage.

Collen Brothers managed to secure a wide variety of contracts not only in Portadown but also throughout the Ulster counties, building private residential dwellings, government buildings and churches. The company built or renovated houses in Portadown and elsewhere, but also undertook more demanding projects, frequently involving a high level of design and technical skill. The firm became the main contractor for public works in their native town. They provided a new town hall for the Portadown Town Commissioners in 1890, delivering an impressive red-brick three-storey building.[18] John Collen also took charge of the construction of a new town hall at Keady, Co. Armagh in 1871, which was built in the Venetian Gothic style, based on the designs of Fitzgerald Louch, an architect from Belfast.[19] The family business also showed considerable ability to secure work from influential private clients. Collen Brothers conducted 'alterations and additions' to Lisbeg House, Ballygawley, Co. Tyrone, in 1871, for George Vesey Stewart, a local landowner and Justice of the Peace.[20] Similarly, the firm secured a contract at the same time 'for the

17. National Archives of Ireland (NAI), Census of Ireland 1911, *Household Return for Thomas Collen*; *Irish Times*, 'Heavy damages against a farmer', 16 February 1906; *Irish Times*, 'Alleged attempt to bribe an architect', 9 June 1909; *Irish Times*, 'Heavy calendar at Belfast', 24 July 1909; *Irish Times*, 'Portadown woman's death: Recent shooting charge recalled', 10 June 1911.

18. *Irish Builder*, vol.32, 'Tenders', 1 February 1890, p.40.

19. *Irish Builder*, vol.13, 'New Town Hall, Keady', 1 June 1871, p.139.

20. *Irish Builder*, vol.13, 'Tenders', 1 November 1871, p.288.

erection of a farmstead' and other works at Ballygawley Park, the country seat of Sir John Marcus Stewart, another substantial landowner in south Tyrone.[21] Collen's private clients were not restricted to the rural gentry, but extended to the banking fraternity as well. The firm provided a new building for the Hibernian Bank in Letterkenny between 1874 and 1875, working with another Belfast architect, Thomas Hevey.[22] Collen Brothers was a flourishing family business within a few years of its establishment, with an impressive client base and a commercial profile that extended throughout Ulster.

Collen Brothers always sought and secured clients across the political and religious divisions that marked late-nineteenth-century Ireland. The firm displayed a particularly ecumenical approach to church building. Collen undertook the construction of a new Methodist church at Banbridge, Co. Down, in 1870, which accommodated about 400 people: the contract also included the provision of a reading room and school at the rear of the new building.[23] It was perhaps not surprising that the firm was providing for the building needs of local Methodist communities, considering the religious heritage of the founding directors, but their involvement in church building ranged far beyond their own religious denomination. The firm was commissioned to construct a new church at Gilford, Co. Down for the Church of Ireland in 1869, while also completing a parsonage house for the resident Anglican clergyman at St Saviour's church in Portadown.[24] Collen built a small church in the Gothic style for the Presbyterian congregation in Portadown during the same year.[25] Moreover the directors were also meeting the building requirements of the Catholic Church, which was then engaged in an intensive phase of church building, less than a decade after the foundation of their business. The firm secured the contract in December 1874 to construct a new Catholic church in Belfast, St Patrick's in Donegall Place; an impressive new building in the Romanesque style, with a spire that rose to a height of 180 feet, was completed three years later.[26] Church building for all the major denominations

---

21. *Ibid.*

22. *Irish Builder*, vol.16, 'Home and Foreign Notes', 15 November 1874, p.306.

23. *Irish Builder*, vol.12, 'Notes of Works', 1 September 1870, p.209.

24. *Irish Builder*, vol.11, 'Notes of Works', 1 December 1869, p.281; Collen Papers Portadown, P. Patterson to Niall Collen, 10 November 1999.

25. *Irish Builder*, vol.11, 'Notes of Works', 15 March 1869, p.71.

26. *Irish Builder*, vol.16, 'Notes of Works', 15 December 1874, p.343; *Irish Builder,* vol.17, 'St Patrick's New R.C. Church, Belfast', 1 January 1875, pp.6–8.

on the island became a significant element of Collen's activity during the late nineteenth century. The approach taken by leading members of the firm to business was shrewd, pragmatic and dictated by commercial priorities.

The company soon expanded well beyond its native region, extending its reach to Dublin in the early 1870s. John and his son Thomas J. Collen (T.J.) established a builders' yard on the banks of the Grand Canal in the city, acquiring land on both sides of the Canal near Mount Street Bridge. The site included a two-storey house at 5 Clanwilliam Place where Collen Brothers opened a small office in 1872, building a new residence alongside the original building.[27] John Collen chose the location carefully: proximity to the canal ensured that barges could easily transport essential raw materials such as sand and gravel from Kildare and limestone from quarries in Tullamore to the builders' yard. Moreover, the central location within the city was a significant advantage at a time when the transport of materials to building sites was entirely by horse and cart.[28] The arrival of Collen Brothers in Clanwilliam Place was marked by a small sculpture, taking the form of a monogram carved on the wall outside the office, which featured the letters 'CB' and incorporated miniature heads of John Collen and his brother Thomas, the two family members who had returned from Australia to take a leading role in the building business.[29] The opening of the company's premises beside the canal marked the beginning of a long-term presence by the Collen family enterprise in Dublin that has endured into the twenty-first century.

Collen operated from Clanwilliam Place for over a generation, maintaining its presence in the city centre until the first decade of the twentieth century. The company acquired land for a new builders' yard on East Wall Road around 1900, initially leasing two and a half acres from Dublin Corporation for a term of 800 years.[30] The construction of a larger builders' yard and offices for the new premises required extensive work, including the building of two new houses at the corner of East Wall Road and the nearby West Road, which was carried out over several years to reduce costs.[31] The new premises provided far greater space and scope for expansion for what was clearly a

---

27. Lyal Collen Papers, *Collen: An Irish Building Family 1782–1992*, p.2.

28. *Ibid.*

29. Lyal Collen, *A Note on Collen Brothers*, p.1.

30. River House Archive, Joseph Collen, 'Telephonic Message', 25 October 1905.

31. *Ibid.*; Henry Overend to Collen Brothers, 11 December 1905; Collen Brothers Ltd. to Mary Ryan, 10 August 1908.

Clanwilliam Place, Dublin, the original offices of
Collen Brothers in the city. Lyal Collen Papers.

growing business; the move to East Wall reflected the successful development
of Collen's operations in Dublin, which soon emerged as a central element of
the company's business. When the company transferred its operations across
the city to the new premises the sculpture duly followed and it currently
remains over the door of the Collen offices in River House.

Collen Brothers acquired several major contracts awarded by public
bodies or non-profit foundations in Dublin during this period. The firm
secured its first project with the Port and Docks Board in October 1881, when
Collen was awarded the contract to build a shed on the North Quay extension.[32] It was a modest beginning to an important commercial relationship,
as the Board would become one of Collen's most important clients almost
a century later in the 1960s. An even more enduring business relationship
was forged in the 1880s, when Collen began a long-term association with the
Royal Dublin Society.

32. Files of Port and Docks Board, 1 March 1882.

The RDS began a gradual move in 1879 from its traditional base in Leinster House to a new site at Ballsbridge. The Society secured a long-term lease (for 500 years) of fifteen acres from the Earl of Pembroke, with the intention of running its agricultural shows on the new site.[33] This meant that a major building programme had to be undertaken in a relatively short time, as the council of the Society hoped to run the Spring Show in Ballsbridge from 1881. The council considered sixteen tenders for the construction of 'a covered hall' at Ballsbridge and in June 1880 Collen Brothers won the contract with a bid of £11,689.[34] The indenture between Collen and the Society, which was formally agreed on 10 July 1880, provided that Collen would 'erect and execute the several buildings and works comprised in this contract in a sound permanent workmanlike and satisfactory manner…'[35] The contract specified that Collen Brothers accepted the responsibility not only for a single agricultural hall but for other necessary building work as well, including new offices for the RDS. The original agreement held the potential for further projects, as the council of the RDS had decided that an extensive building programme in Ballsbridge would be implemented over several years.[36]

Collen Brothers took full advantage of the considerable opportunity offered by the contract. The company built the new Main Hall, with an area of around 60,000 square feet, and front offices for the Society, all within seven months.[37] While this project was still in progress, the council of the RDS asked Collen Brothers in November or December 1880 to remove the Agricultural Hall in Kildare St and rebuild it at Ballsbridge, without requiring them to submit a new tender.[38] The Irish administration agreed to allow materials from the old Agricultural Hall to be re-used in its replacement structure at Ballsbridge, while the RDS also secured a grant from the Board of Works to cover the cost of relocating the building.[39] The council moved swiftly to use this unexpected windfall by instructing Collen to undertake the immediate

---

33. *The Royal Dublin Society 1731–1981*, ed. Desmond Clarke and James Meenan (Dublin, 1981), p.71.

34. *Irish Builder*, vol.22, 'The Royal Dublin Society and the Science and Art Department', 15 June 1880, p.168.

35. River House Archive, *Indenture between Collen Brothers and the Royal Dublin Society*, 10 July 1880.

36. *Irish Times*, 'Royal Dublin Society', 4 June 1880.

37. River House Archive, *Royal Dublin Society*, p.1.

38. *Ibid.*; *Irish Times*, 25 August 1900.

39. *Irish Times*, 'Royal Dublin Society', 10 December 1880.

transfer of the building. This approach attracted some criticism at a meeting of the Society on 10 December 1880, but Samuel F. Adair, a member of the council, defended the decision on the basis that the project involved only the transfer of the old building to Ballsbridge as an addition to the work already undertaken by Collen. Another member of the Council, Lt-Colonel Vesey, commented that the decision avoided 'any unpleasantness arising in case a fresh contractor was engaged to do the work', as the company already had the right under the contract to construct all the buildings at Ballsbridge and would insist on erecting the Agricultural Hall on the new site even if another contractor was allowed to remove it. Moreover Vesey commented that 'Messrs Collins [sic] had given them every satisfaction in the manner in which they were carrying out the contract.'[40] The advocates of employing a single contractor won the day and the transfer of the Agricultural Hall by Collen went ahead early in 1881 at a cost of £3259. The reconstructed building became known as the South Hall on the new site and its galleries were placed around the Main Hall.[41] The major building projects were completed by June 1881, enabling the Society to hold both of its flagship events, the Spring Show and the Horse Show, at Ballsbridge for the first time.

The RDS undertook a gradual process of expansion on the new site, creating additional opportunities for the Collen business. The firm extended the new stand and added ancillary facilities for staff at Ballsbridge in 1882.[42] The company secured further building projects from the Society during the 1890s. They built a new two-storey building, designed by the architect C.H. Ashworth, on both sides of the South Hall in 1896.[43] The Society proved willing to offer the company repeat business on a regular basis, underlining the council's preference for a single contractor developing the Ballsbridge premises, as well as signalling its approval of the building work already undertaken by the firm. Collen played a central part in the successful evolution of the RDS at Ballsbridge during the final two decades of the nineteenth century. The firm's professionalism in delivering its side of the bargain, with efficient and cost-effective completion of the building projects, did a great deal to ensure

40. *Irish Times*, 'Royal Dublin Society', 10 December 1880.

41. *The Royal Dublin Society 1731–1981*, ed. Desmond Clarke and James Meenan (Dublin, 1981), p.71; Terence de Vere White, *The Story of the Royal Dublin Society* (Tralee, 1955), pp.159–60; H.F. Berry, *A History of the Royal Dublin Society* (London, 1915) p.312.

42. *Irish Builder*, vol.24, 'The Royal Dublin Society's Show-Yard', 1 September 1882, p.249.

43. *The Royal Dublin Society 1731–1981*, ed. Desmond Clarke and James Meenan (Dublin, 1981), p.72.

that the Society became one of Collen's most important long-term clients. The success of the early collaboration in the 1880s forged an enduring association between Collen Brothers and the RDS, which would prove immensely valuable to both parties throughout the following century.

Collen Brothers was increasingly successful in winning contracts throughout Ireland, from the building of Furlong's Mills in Cork to Renmore Barracks in Galway city. The contract for Renmore Barracks was acquired from an important new client – the British army. Collen completed the new barracks between 1876 and 1880.[44] The construction of Renmore barracks underlined the expansion of the company, which was now able to complete major projects at a considerable distance from its original base in Portadown or even its latest headquarters in Dublin. The firm later secured valuable contracts at the Curragh military camp and the Royal Hibernian Military School in the first decade of the twentieth century.[45] The contract for Renmore marked the beginning of a business relationship with the British army in Ireland, which would continue through the First World War and would be maintained by the northern branch of the family even after the partition of the island in 1921–22.

Yet the firm was not simply dependent on its contracts with the British state or prominent social institutions such as the RDS. Collen Brothers also undertook several major projects initiated by private clients, often drawn from the Anglo-Irish aristocracy, shortly after the establishment of the firm's offices in Dublin. These contracts included the rebuilding of Kylemore Abbey in Galway for the Duke of Manchester and the construction of Killarney House for the fourth Earl of Kenmare. Although the company was not initially involved in the construction of Kylemore Castle, the purchase of the mansion by the Duke of Manchester and his dissatisfaction with the design of its interior opened up a valuable business opportunity for Collen, not least because the Duke was also the owner of Tandragee Castle. The contractors from Portadown enjoyed a strategic advantage in bidding for the contract and it was not surprising that Collen duly undertook the internal reconstruction of Kylemore Castle.[46] The building was later sold once again, this time to a Belgian order of nuns, and its name was changed to Kylemore Abbey.

The building of Killarney House proved a very different and much more demanding undertaking – it was the most extensive project undertaken by

---

44. Lyal Collen Papers, *The Collen Building Tradition*, p.1..

45. See Chapter 2.

46. Lyal Collen Papers, Lyal Collen Note, *Kylemore Abbey*, p.1.

Killarney House, Kenmare. Lyal Collen Papers.

the firm for a private client in this period. The Earl and Countess of Kenmare decided in 1872 to replace their existing family home in south Kerry with a much more impressive aristocratic residence on a nearby site.[47] The new house was designed by an English architect, George Devey, although the work was supervised by William Henry Lynn of Belfast; Collen Brothers secured the building contract. The old house was demolished and a vast red-brick mansion in the Tudor and Victorian style was built on a hilltop north-west of the original house.[48] Joseph Collen (1856–1941), John's younger brother who began to work for the firm in his early twenties, took charge of overseeing the project. He showed considerable dedication to the task, renting a house in Killarney where he lived with his young family, so as to manage the job more effectively.[49] The building work was completed between 1877 and 1878. The construction of Killarney House demanded a high level of building and architectural expertise. The mansion included a series of gables, some triangular and others curvilinear in shape; it also featured a number of decorative oriels in different shapes. The interior of the residence was panelled or hung with Spanish leather and adorned with carved wooden columns. Mark Bence-Jones

47. Mark Bence-Jones, *A Guide to Irish Country Houses* (London, 1996), p.162.

48. *Ibid.*

49. Lyal Collen Papers, *The Collen Building Tradition*, p.2.

in his *Guide to Irish Country Houses* described Killarney House as 'one of the wonders of Ireland'.[50] Lyal Collen reflected the more pragmatic thinking of his family – or the institutional memory of how difficult the job had proved to be – when he referred to the mansion as a 'magnificent but over elaborate building'.[51] The imposing edifice did not survive for longer than a generation. Killarney House was gutted by fire in 1913 and never rebuilt on its original site.[52] Yet the building of such a vast and architecturally elaborate building, with a variety of intricate decorative features, was a significant achievement for a relatively new construction company.

The firm also enhanced its profile in a more traditional area of its operations, church building, during the 1880s. Collen made a successful tender for the construction of a new church in Raheny for the local Church of Ireland congregation. The All Saints church was financed by Lord Ardilaun, a leading member of the Guinness family, who wished to endow the church in commemoration of his father.[53] The building itself was designed in the Gothic style by the architect George C. Ashlin and attracted favourable attention from *The Irish Builder*, which praised the architect's willingness to embellish the church with distinctive stonework carvings:

> …we ought to be thankful to the principal or architect who affords us a little variety or novelty in the way of handicraft, just to relieve us from the poverty and monotony of the moulding-machine dead-face work which has well-nigh left us in latter times with potatoes and milk jobs, as well as ditto for dinner all the year round![54]

Ashlin's ambitious design and preference for intricate stonework certainly produced an impressive ecclesiastical structure, although it also imposed greater demands on Collen and slowed the pace of the building work. The firm completed the project over a four-year period, starting in September 1885 and concluding towards the end of 1889.[55] Collen's work found favour with the leading newspaper of the Protestant political and religious establishment. The *Irish Times* commented approvingly in December 1889 that 'there can be no two opinions as to the superior style in which the contractors have discharged the task allotted to them, resulting in the erection of one of the prettiest

---

50. Mark Bence-Jones, *A Guide to Irish Country Houses* (London, 1996), p.162.

51. Lyal Collen Papers, *The Collen Building Tradition*, p.2.

52. Mark Bence-Jones, *A Guide to Irish Country Houses* (London, 1996), p.162.

53. *Irish Times*, 'All Saints' Church, Raheny', 14 December 1889.

54. *Irish Builder*, vol.28, 'New Church, Raheny, County Dublin', 15 May 1886, pp.150–151.

55. *Irish Times*, 'All Saints' Church, Raheny', 14 December 1889.

churches in Ireland'.[56] The imprimatur of the *Irish Times* certainly illustrated Collen's increasing reputation in the construction business; the article and the contract itself also reflected the firm's success in forging connections with influential figures and institutions within the traditional Protestant upper class, from Lord Ardilaun to the Royal Dublin Society.

Shortly before the turn of the century, the firm's growing reputation enabled Collen to secure its most significant contract yet. The Board of Control of lunatic asylums in Ireland sought tenders in February 1896 for the construction of a new asylum for patients with mental disabilities at Portrane, in Co. Dublin.[57] Collen Brothers won the contract on 6 August 1896: the new asylum was to be completed in 1902 at a price of £167,000.[58] The decision by the Board of Control to award the contract to Collen attracted some controversy, ostensibly due to the firm's origins in the north of Ireland. When the Governors of Richmond District asylum were notified of the Board's decision in August 1896, at least one member raised strong objections; William J. Moore alleged that Collen's tender was higher than one of its competitors and complained that 'it was a pity the contract was not given to a Dublin firm'.[59] Dr John Eustace, another member of the Governors, defended the decision on the basis that at least the contract had been given to an Irish firm. The Governors agreed to seek details from the Board about the tenders submitted earlier in the year, but their ability to raise any objection was limited, as the power to award the contract rested with the Board of Control in the Custom House.[60] The complaint raised had no practical impact and Collen commenced its operations on the site later in the same year. Yet it was striking that Collen could still plausibly be presented as an interloper in the commercial life of the city, although the firm had established itself in Dublin over two decades previously. There is little doubt that the complaint was motivated by resentment within an element of the local commercial elite at the rapid success enjoyed by the brothers from Portadown.

The new asylum at Portrane was urgently required as the existing facility at Richmond was seriously overcrowded, accommodating over 1700 patients in a

---

56. *Ibid.*

57. *Irish Builder*, vol.38, 'New Lunatic Asylum, Portrane, Co. Dublin', 1 August 1896, p.157.

58. Collen Papers Portadown, Secretary of the Office of Control of Lunatic Asylums to Collen Brothers Ltd., 6 August 1896.

59. *Irish Builder*, vol.38, 'New Lunatic Asylum, Portrane, Co. Dublin', 1 August 1896, p.157.

60. *Ibid.*; *Irish Times*, 14 August 1896.

building that was intended to hold no more than 1100.[61] The new asylum was designed to accommodate 1200 patients, divided between different wings to hold patients with different forms of disability.[62] The buildings for the patients were constructed so that each floor contained one or more self-contained wards, avoiding any necessity to use stairs within the wards. The company built an administration block, the dispensary, visitors' rooms and a large dining hall, which was also equipped for dramatic productions. The new building included Catholic and Protestant chapels, which may have been designed as much to divert the unfortunate inmates as to provide for their spiritual salvation. The company also constructed an impressive loft clock tower at a height of 130 feet, which contained a large water tank for use in the event of a fire. Collen Brothers made substantial progress with the building work by 1900, when *The Irish Builder* reported that almost all the buildings had been roofed in and several were close to completion.[63] The company completed the permanent buildings, drainage work and paths for the new asylum by 1903, apparently bringing its involvement in the project to a satisfactory conclusion.

The workforce at Portrane, c.1900; Harky Collen is third from left. Lyal Collen Papers.

61. *Irish Times*, 14 August 1896.

62. *Irish Builder*, 'Our Supplement', 15 July 1900, pp.420–1.

63. *Ibid.*

Yet tensions arose between the company and the architect in charge of the project, George C. Ashlin, over the costs of the work. Collen Brothers complained to Ashlin that the quantity surveyors were failing to take account of extra works and deviations from the original contract that were regarded as necessary by the company. Collen wrote to the architect on 17 October 1901 to complain that the surveyors 'are so very slow and seem to suit their own convenience only in dealing with the deviations.'[64] This complaint fore-shadowed a protracted legal dispute. Collen submitted their final account for payment to the architect in June 1903. The company's bill under the original contract, amounting to over £177,445, was paid in a timely fashion by the Portrane asylum committee, representing the board of Governors for Richmond asylum.[65] But Collen's claim for extras, which involved work done outside the terms of the original contract at the instigation of the board, was a very different matter: the company's assessment of its costs diverged widely from Ashlin's much more conservative estimates, leaving the sum of over £27,000 in dispute between the board and Collen Brothers.[66] It was soon apparent that no agreement would be possible and Collen referred the dispute to arbitration in November 1903.[67] The two parties agreed that Sir Thomas Drew would act as sole arbitrator. The arbitration hearings began on 14 March 1905, continuing for thirty-five days. The eventual outcome was a compromise that offered something to each side. Collen was awarded £10,108 as an additional payment in settlement of the claim; the board was able to conclude the dispute at a significantly lower price than the company's original demand.[68] While the board was relieved at the decision, the award itself vindicated Collen's decision to pursue the case and upheld at least in part the company's refusal to accept the architect's estimates. The dispute underlined that John Collen, still the leading figure in the company, had lost none of his willingness to confront architects or even influential clients where he considered their approach financially unreasonable or unfair to his firm. This combination of financial prudence and tough-minded negotiation would remain key features of Collen business practice throughout the following century.

---

64. River House Archive, Collen Brothers to G.C Ashlin, 17 October 1901.

65. Report of the Sub-Committee for Richmond Board of Governors, *Collen Brothers, Contractors v. Portrane Asylum Committee – Sir Thomas Drew's award*, 9 September 1905.

66. *Irish Times*, 'Portrane Asylum Arbitration', 12 April 1905.

67. Report of the Sub-Committee for Richmond Board of Governors, *Collen Brothers, Contractors v. Portrane Asylum Committee – Sir Thomas Drew's award*, 9 September 1905.

68. *Irish Times*, 'Richmond Lunatic Asylum', 1 September 1905.

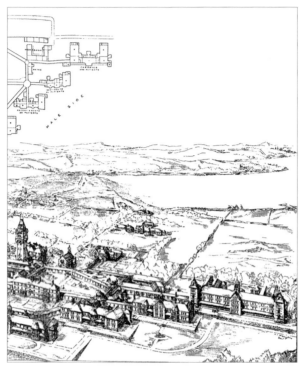

Portrane Asylum. Lyal Collen Papers

The construction of Portrane asylum was a vast project that had lasted almost seven years and demanded substantial resources of workers, building materials and capital. The building of the new asylum was not only the most significant project undertaken by Collen in its first generation, but also the largest building contract ever awarded to a single contractor in Ireland before 1900.[69] The total price paid by the board for the company's services, following the award in 1905, was over £203,000.[70] It is evident that much of this payment was absorbed by the company's costs and the level of profit on the contract remains unknown, but Collen certainly benefited handsomely from such a major project. The award and successful implementation of the contract reflected the established position secured by Collen within the Irish building sector and the company's rapid expansion from its modest origins in Portadown.

---

69. *The Irish Builder*, 'Our Supplement', 15 July 1900, pp.420–1.

70. Report of the Sub-Committee for Richmond Board of Governors, *Collen Brothers, Contractors v. Portrane Asylum Committee – Sir Thomas Drew's award*, 9 September 1905.

While the firm's activity in its first generation was focused primarily on building projects, Collen Brothers also enjoyed a relatively brief but productive venture in railway construction. Collen issued a successful tender for the construction of the Cavan and Leitrim railway in 1885. The firm was given the task of laying down the track and bridges for the railway, which would extend for thirty-two miles from Belturbet to Dromod.[71] It was a large-scale civil engineering project, involving excavation of soil, preparation of the necessary formation, laying the sleepers and rails and the construction of bridges. The excavation was a laborious process, carried out by large groups of workers wielding picks and shovels – steam shovels were being pioneered in Britain around this time, but had not yet been introduced in Ireland.[72] The workers used rail wagons pulled by horses to deposit the excavated material on the line of the track. Collen utilized steam locomotives on a three-foot gauge to transport building materials for longer journeys, but the two locomotives employed by the firm, the *Express* and the *Victor*, had been acquired second hand and their reliability proved dubious. The use of the locomotives was unavoidable due to the need to move heavy loads of building materials to the construction site, but it occasionally had tragic consequences. A serious accident occurred on the evening of 15 March 1887, when a train carrying labourers home from work struck a donkey near the railway crossing at Clooncahir. All the wagons were derailed and three workers were killed in the crash, while four others were injured.[73] The tragedy was caused not by the unreliability of the steam engines, but by a mistake on the part of the driver, as the train was travelling at an excessive speed without lights of any kind. This accident provoked a legal case against Collen and the railway company. The courts subsequently gave a verdict against the building contractor on the basis that the careless fashion in which the train was driven amounted to neglect, justifying an award of compensation to the victims and their families.[74] The train crash was the worst accident affecting Collen Brothers' employees since its foundation and one of the most severe in the history of the company.

Despite the tragic accident and subsequent court proceedings, the Cavan and Leitrim railway was completed by Collen within three years. The track was laid by the beginning of July 1887 and the new railway saw its first train

---

71. Lyal Collen Papers, Lyal Collen Note, *Cavan and Leitrim light railway*, p.1.

72. *Ibid.* p.3.

73. P.J Flanagan, *The Cavan and Leitrim Railway* (Newton Abbot, Devon, 1966), p.19.

74. *Ibid.*

on 26 July, although work continued on other aspects of the project for almost another year. The project was not fully completed until May 1888, when the line was opened to all traffic.[75] The firm also acquired an additional contract in the same area, the construction of a narrow-gauge tramway from Ballinamore to Arigna, which was intended to serve the coal mines in Arigna.[76] On this occasion Collen became embroiled in a brief but violent industrial dispute with a group of its workers. The labourers objected in December 1886 to the proposed line of the tramway, which would cross the green where Ballinamore Fair was usually held. A strike broke out, followed by violent scuffles between the management and labourers and the arrest of several workers.[77] This was not a case of industrial action by workers seeking improved pay and conditions, which was already becoming a regular feature of the building industry in Ireland, but a militant protest by local labourers against an apparent intrusion on a traditional fair-ground. The company responded in an equally militant fashion, as Lyal Collen acknowledged in his notes on the early history of the firm: 'The problem was resolved when thirty seven Dublin men were recruited and camped out on the site in large tents.'[78] The firm recruited labourers from Dublin to work on the line and they set up camp at night in 'a semi military' fashion.[79] The newly imported labourers were the focus of considerable resentment from local workers; a force of police guarded their camp at night and patrolled the construction site during the day.[80] From our present vantage point in the early twenty-first century, such a confrontational tactic by an employer would appear more likely to exacerbate the dispute than ameliorate it. Yet the protest and Collen's response should be understood in the context of the time. The late 1880s were marked by bitter agrarian agitation, as nationalist politicians and tenant farmers challenged the British authorities with a sustained campaign for land reform. Moreover the evolution of labour law and the regulation of labour disputes by the state lay some way in the future; militant tactics by either party to a dispute were commonplace. In any event the clash between management and labourers at Ballinamore was intense and violent, but short-lived. The local labourers returned to work in January 1887 after a strike of one

---

75. *Ibid.* pp.21–2.

76. Lyal Collen Papers, Lyal Collen Note, *Cavan and Leitrim light railway*, pp.3–4.

77. P.J. Flanagan, *The Cavan and Leitrim Railway* (Newton Abbot, Devon, 1966), p.20.

78. Lyal Collen Papers, Lyal Collen Note, *Cavan and Leitrim light railway*, p.4.

79. *Ibid.*

80. P.J. Flanagan, *The Cavan and Leitrim Railway* (Newton Abbot, Devon, 1966), p.20.

week, almost certainly because they could not afford to stay out any longer.[81] There were no further industrial disputes and the construction of the tramway was completed despite some delay caused by the protest. The dispute did not deter Collen from tendering for other narrow-gauge railway projects, but the firm succeeded in winning only one further contract, laying the line from Dundalk to Carrickmacross.[82] The venture into railway construction proved a successful exception to the normal rule of Collen's building activity in the late nineteenth century rather than the prelude to a more sustained engagement with large-scale civil engineering projects.

Railway Bridge, Cavan and Leitrim light railway, *c*.1886. Courtesy of Collen Construction.

The firm, however, acquired several other contracts as a result of the boom in railway construction in nineteenth-century Ireland, mainly involving the construction of hotels serving rail passengers. When the Midland and Great Western railway line was extended in 1894 to Westport and Achill Island in Co. Mayo, the expansion of the rail network created new opportunities for

---

81. *Ibid.* pp.20–1.

82. Lyal Collen Papers, Lyal Collen Note, *Cavan and Leitrim light railway*, p.4.

Collen. The firm secured a contract in 1895 with the Great Southern and Western Railway Company to build a new railway hotel at Mulranny, near Westport, working from the designs of Sir Thomas Deane & Son, a well-known architectural practice based in Dublin.[83] Collen adopted the practice of giving the direction of major contracts to family members, who worked in conjunction with experienced foremen. The project at Mulranny was no exception, providing the first experience of management to one of the younger members of the family: Joseph Harcourt Collen (known as Harky), who was Joseph's younger son, took a leading role in overseeing the job.[84] The contract proved a baptism of fire for the eighteen-year-old Harky, not least because the foreman on the site was an alcoholic, ensuring that the younger man had to assume greater managerial responsibility. Harky lived at Mulranny for two years, following the example set by his father in Killarney twenty years earlier. It was not all work though: Harky, who had a keen interest in horseracing, took the opportunity to engage in pony racing competitions on the beaches of Mulranny, Achill and Belmullet.[85] The job itself combined demanding responsibilities in managing the site, with moments of excitement and occasional danger. Harky collected the wages for the workers, which were delivered by train to Newport every month. As the writ of the Royal Irish Constabulary did not always run to more remote areas of the country, it was a potentially dangerous undertaking. Harky made the journey on one of his racing ponies, carrying a Smith and Wesson revolver for his own protection and the safety of the wage packets.[86] The firm's workers also faced the more mundane challenge of removing large rock outcrops on the site with explosives. Collen completed the new hotel in 1897 at a price of £7686.[87] The firm also took charge of other ancillary projects serving newly built railway lines in the west of Ireland towards the end of the nineteenth century, constructing another railway hotel in Westport and a new station house at Dromod for the Midland and Great Western line. The Mulranny project, however, established an enduring resonance in Collen family tradition, receiving favourable attention in Lyal

---

83. IAA, Patterson, Kempster and Shortall (PKS), 77/1/A07, Cash account book, entries for June 1895 and March 1896, p.167.

84. Lyal Collen Papers, Lyal Collen Note, *Malaranney (sic)*, p.1.

85. *Ibid*.

86. *Ibid*.

87. IAA, PKS, 77/1/B18/07, Letter book, *Bill of Measurement of Extras and Deviations at Mullarany Hotel* (sic), 22 November 1897, pp.164–99; *Summary*, p.213.

Collen's notes on the business almost a hundred years later.[88] While the project presented some unusual challenges and was undoubtedly more demanding than other contracts of the same kind, it was primarily Harky's adventures on the western seaboard that guaranteed Mulranny a permanent place in the collective memory of the family.

Mulranny Bay Hotel, Co. Mayo. Lyal Collen Papers.

The firm adopted a recognizably modern corporate structure for the first time in the final decade of the nineteenth century. The partners agreed in June 1894 that the firm of Collen Brothers should be reconstituted as a limited liability company.[89] Collen Brothers Ltd came into existence formally on 21 January 1895.[90] The agreement on 16 June 1894 was concluded by John Collen, his son T.J. and his brothers Joseph, Richard and David. The five signatories agreed to invest in the business and received shares in proportion to their stake: the total capital held by the company was £30,000, which was

88. Lyal Collen Papers, Lyal Collen Note, *Malaranney (sic)*, p.1.

89. River House Archive, *Proposition to change the Firm of Collen Bros into a Limited Liability Co. as follows*, 16 June 1894.

90. Ministry of Commerce, Government of Northern Ireland, Certificate under the Companies (Reconstitution of Records) Act, Northern Ireland, 1923, for Collen Brothers Ltd (with original date of incorporation), 25 September 1923; Speech by Joe Collen to the Rotary Society, *My Job*, March 1971, p.1.

composed of 300 shares at a value of £100 per share. Most of the original share capital – £24,000 – consisted of ordinary shares taken up by members of the family, while preference shares accounted for a further £6000. The agreement provided that 'each active or working partner' would receive a share of the profits in proportion to their capital investment.[91] It is evident that all five shareholders were expected to take an active part in the running of the business, if they wished to secure a share of the profits. This did not imply that all the partners had an equal stake in the newly reformed company – in fact the investment made by the original directors varied widely. John Collen was the largest shareholder, taking up a hundred shares and providing a third of the company's share capital; his younger brother Joseph took up the next highest stake of sixty shares. T.J. and Richard enjoyed the more modest holdings of thirty shares each, while David took up the lowest stake of twenty. This distribution reflected the reality that some partners were undoubtedly more influential than others.

It was hardly surprising that John Collen, the original founder of the business and still its most dominant figure, was also the largest shareholder. Similarly, Joseph was a very active and successful manager within the firm by 1894, having supervised the construction of Killarney House. The two men were more deeply involved in the management of the company than their siblings or relatives. The original agreement on 16 June 1894 provided that Joseph was to be paid an additional £100 annually, which would be charged to trade expenses.[92] This extra payment recognised Joseph's managerial responsibilities and his increasingly prominent role in the business. The founding agreement also stipulated that John and Joseph were 'to sell or give one or more shares to some member of each of their [own] family' to ensure that the required number of shares were fully allocated: this provision did not apply to the other shareholders.[93] The inclusion of this clause had significant implications for the long-term future of the business, as it indicated strongly that the successors to the original partners would be found among the children of the most active directors. Joseph's sons, John Black and Harky, became active in the company for the first time in the 1890s – ultimately it would fall to them to take over the management of the business.

---

91. River House Archive, *Proposition to change the Firm of Collen Bros into a Limited Liability Co. as follows*, 16 June 1894.

92. *Ibid.*

93. *Ibid.*

Yet whatever the distribution of influence between the partners, the most striking feature of the agreement was the extent to which Collen Brothers Ltd was above all a family business. The new limited company was wholly owned, financed and managed by members of the family. The central place of the family within the business was the most influential legacy of John Collen and his brothers, and it would remain the defining feature of the company over the following century. The founding generation could claim an impressive record of achievement. The transformation of the original family firm into a stable limited company was a significant milestone, but by no means the sole accomplishment of the founding partners. Collen Brothers Ltd was firmly established as a flourishing construction business before the end of the nineteenth century. The company operated on an all-Ireland basis, expanding with remarkable speed from its initial base in Portadown and winning major contracts in all four provinces. Collen soon developed strong connections with influential individuals and institutions within the largely Protestant social and cultural establishment of late-nineteenth-century Ireland. Yet Collen's business practice was characterized by shrewd pragmatism and was not significantly influenced by political or religious concerns: a key strength of the company was its ability to attract and maintain a wide variety of clients, from the British army to the Royal Dublin Society to the Catholic Church. The company's pragmatism and resilience would be fully tested in the following generation, when extraordinary political and social change swept away many of the familiar landmarks of the previous era.

TWO

War, Partition and Survival

The partners of Collen Brothers at the beginning of the twentieth century were not only prosperous building contractors but also established members of the social elite in Victorian and Edwardian Ireland. The status of the Collen family had been transformed over the previous half-century. The career of John Collen underlined the dramatic change in the family's fortunes. The young stonemason who had emigrated to Australia in the 1850s was a wealthy businessman, magistrate and local political figure by the first decade of the twentieth century. While he was often in Dublin supervising the activity of the company, he also acquired a local residence at Killycomain House near Portadown, which became his family home.[1] John Collen enjoyed considerable prestige and status in his native town, playing a prominent part in the civic life of the local region. He was appointed as a JP for Portadown and emerged as a leading member of the local Unionist Party in north Armagh. He was present in January 1894 at a political rally in Portadown, which was addressed by the Marquis of Londonderry, a unionist aristocrat and strong opponent of home rule for Ireland.[2] This was not an isolated episode – John Collen's

1. NAI, Census of Ireland 1911, *Household Return for John Collen.*

2. *Irish Times*, 'Great Unionist meeting in Portadown: Important speech by Lord Londonderry', 25 January 1894.

commitment to the Unionist Party remained a consistent theme for the rest of his life. He was a member of the General Committee of the North Armagh Unionist Association during the first decade of the twentieth century and was present in December 1909 to support the reselection of the sitting Unionist MP, William Moore, to contest the general election of January 1910.[3] While John's political activism always took second place to his overriding preoccupation with business, he undoubtedly had a strong interest in national political issues and was personally involved in local politics; following the passage of the Local Government Act in 1898, he was selected to represent Portadown Urban District as a member of Armagh County Council.[4] He remained a member of the predominantly Unionist council for over two decades until his retirement due to ill health in 1920.[5] John Collen's firm attachment to the crown and the union between Britain and Ireland were evident during his tenure on the council, not least when he seconded the motion in June 1903 to present a loyal address from the council to King Edward VII and Queen Alexandra on their first royal visit to Ireland.[6]

John Collen, pictured as Deputy Lieutenant and
High Sheriff of Armagh, 1911. Courtesy of Niall Collen.

3. *Irish Times*, 'Mr William Moore Again Selected', 16 December 1909.

4. IAA, *Dictionary of Irish Architects*, John Collen 1837–1921.

5. *The Irish Builder*, vol.63, 'Obituary: Mr. John Collen D.L., Portadown', 4 June 1921, p.390.

6. *Irish Times*, 'Armagh County Council', 18 June 1903.

The founder of Collen Brothers accumulated further prestigious offices in the early 1900s. He was appointed as a Deputy Lieutenant for the county in December 1906, retaining the post until his death. He also served as High Sheriff of Armagh in 1911, attending the coronation of King George V and Queen Mary.[7] John Collen was a pillar of the Unionist establishment in Portadown, although he was arguably more of a magistrate than a politician and more of a businessman than either. His appointment as a Deputy Lieu-tenant of Armagh occurred under a Liberal government and was confirmed by Lord Aberdeen, a Liberal peer and supporter of Irish home rule who was appointed as Lord Lieutenant of Ireland in 1906. The timing of Collen's appointment suggests that he owed his elevation more to his prominent status as a leading employer and local public figure in Portadown, than to his undoubted connections with the Unionist Party. It was significant too that John Collen's Unionist political orientation did not affect the business practice of Collen Brothers. The company employed foremen and labourers without regard to their religion and maintained an eclectic range of clients from the British army in the Curragh to the Catholic Church in Belfast.

While few of his relatives shared either John Collen's prominent position in his native region or his willingness to engage in local politics, almost all enjoyed a comparable level of affluence, largely due to the success of Collen Brothers. Richard, one of the founding partners of the firm, was able to buy Glenmore House in Rostrevor, Co. Down, which was later described by the *Irish Times* as 'one of the prettiest residences on Carlingford Lough'.[8] David Collen settled in Dublin, acquiring a house and farm in Kilbarrack – he was described on the 1911 census of Ireland as a 'building and engineering works contractor and farmer.'[9] As the census suggests, David gave much of his attention to farming: he also displayed a keen interest in show-jumping competitions at the RDS, owning horses which competed at Ballsbridge with varying degrees of success.[10] He later became deeply engaged in the affairs of the Methodist church, serving as a prominent member of its standing committee in Ireland during the interwar period.[11] David was involved in the

7. *Irish Times*, 'New Deputy Lieutenants', 19 December 1906; *Irish Times*, 'Mr. John Collen, D.L.' 21 May 1921.

8. *Irish Times*, 'Mr. Richard Collen', 8 February 1927.

9. NAI, Census of Ireland 1911, *Household Return for David Collen*.

10. *Irish Times*, 'Jumping at the Show', 4 August 1932.

11. *Irish Times*, 'The Methodist church: Meeting of Standing Committee', 15 October 1935;

inspection of church property, giving advice on the renovation of churches in various parts of the country and was described by the *Irish Times* after his death in 1945 as 'the Committee's honorary adviser on building matters'.[12] It was Joseph Collen who emerged as the most active and dominant figure within the firm during the first two decades of the twentieth century, not least because John was now elderly and based mainly in Portadown. Joseph sold his house at Woodbrook in Armagh to Collen Brothers as part of the agreement establishing the company in 1894 and moved to Dublin, securing a spacious six-room house in Balally, Dundrum.[13] He took the lead in steering Collen Brothers through the First World War and the subsequent military conflict between the British state and the movement for national independence.

Joseph Collen. Lyal Collen Papers

Irish Times, 'The Methodist church: Representatives to Conference – Meeting of General Committee', 20 May 1937; *Irish Times*, 'Methodist church', 13 March 1944.

12. *Irish Times*, 'Obituary: Mr. David Collen', 6 February 1945.

13. NAI, Census of Ireland 1911, *Household Return for Joseph Collen*; River House Archive, *Proposition to change the Firm of Collen Bros into a Limited Liability Co. as follows*, 16 June 1894.

It is apparent that most male members of the family gravitated towards the construction business in general and Collen Brothers in particular. John Collen's second son William proved a notable exception, developing his own career as a highly qualified engineer in the service of the British administration in Ireland. Following a successful apprenticeship as an assistant engineer, including a period working on the Cavan and Leitrim railway, William Collen was appointed County Surveyor for Dublin in 1897, taking responsibility for all the engineering business of the county.[14] He pioneered new techniques of road maintenance, introducing steamrolling to southern Ireland for the first time. William Collen was a consistent advocate of the value of steamrolling as the most effective and economic method of road maintenance and his position gave him the opportunity to put his views into practice.[15] His innovative approach brought about a marked improvement in the condition of the roads in Co. Dublin.[16] William's distinguished career in the public service underlined the extent to which the Collen family had become engaged in the social and commercial life of Dublin by the first decade of the twentieth century. John Collen and his family had acquired a considerable stake in the existing political and social order, which would soon be threatened by war, revolution and civil strife.

The commercial success enjoyed by Collen Brothers in the late 1800s was sustained in the early years of the new century. The company acquired its fair share of major contracts both in Dublin and its traditional hinterland in Ulster. Collen provided an impressive new building for the North British and Mercantile Insurance company between 1900 and 1902; the new offices were situated on the site where Morrison's Hotel had previously stood, at the corner of Dawson Street and Nassau Street in Dublin's city centre. The building was designed in the Renaissance style and built with Irish limestone; it was surmounted by a dome and cupola finished in copper.[17] The walls of the entrance hall were lined with red and white marble from Cork, Connemara and Sicily, while the offices inside the building featured a pattern of black and white marble.[18] The *Irish Times* waxed lyrical in praising the outcome,

14. IAA, *Dictionary of Irish Architects*, William Collen 1861–1932; Interview with Dr Ronald Cox, 30 July 2009.

15. *Ibid.*; Joseph Mooney, 'The Advantages and Economy of Maintaining Good Roads: Discussion', *The First Irish Roads Congress, Records of Proceedings* (Dublin, 1910), pp.25–6.

16. IAA, *Dictionary of Irish Architects*, William Collen 1861–1932.

17. *Irish Times*, 'North British and Mercantile Insurance Company', 9 August 1902.

18. *Irish Builder*, vol.43, 'Topical Touches', 31 July 1902, p.1335.

describing the insurance company's new home even before its completion as 'one of the finest buildings in the city, both as regards extent and from an architectural point of view.'[19] *The Irish Builder* was more prosaic, commenting that the works 'are certainly a credit to the contractors', but also gave the new building a favourable reception.[20]

Shortly afterwards Collen secured a less architecturally imposing but equally valuable project close to the firm's original base, winning a contract to implement the Portadown and Banbridge water scheme in 1904.[21] The water supply for Portadown at the beginning of twentieth century was still obtained from wells and springs, supplemented by the storage of rainwater in barrels or tanks. The local authorities for Portadown and Banbridge came together in 1902 to form a joint Board, which set out to establish a public water supply for the two towns.[22] The scheme devised by the Board's engineers involved piping water from two rivers in the Mourne mountains – this plan required the construction of a storage reservoir, a dam across the valley at Fofanny to impound the water, a service reservoir and pipes to transport the water to Portadown and Banbridge. The Board divided the project between two contractors in March 1904 – Collen Brothers was awarded the contract to provide the service reservoir and pipe work, which was valued at £36,000, while the building of the dam and the storage reservoir at Fofanny were allocated to John Graham of Dromore.[23] Collen was entrusted with the more onerous and time-consuming task, involving the installation of a piped water supply for the two towns. The water was distributed to Portadown and Banbridge by gravity feed through 9- and 10-inch cast iron pipes. This was the most protracted element of the project, which was completed over a five-year period: the piped water supply was distributed throughout Portadown by 1909.[24] The installation of the public water supply for the two towns was a large-scale project, illustrating the key role played by Collen Brothers in providing public works schemes for local authorities in Armagh and Down during this period.

---

19. *Irish Times*, 'North British and Mercantile Insurance Company', 9 August 1902.

20. *Irish Builder*, vol.43, 'Topical Touches', 31 July 1902, p.1335.

21. *Irish Times*, 'Belfast and the North: A large contract', 30 March 1904.

22. S.C Lutton, 'Portadown and Banbridge Regional Waterworks – Foffany Catchment Area', in *Review: Journal of the Craigavon Historical Society*, vol.1, no.1 (1969).

23. *Ibid.*

24. *Ibid.*

Collen Brothers developed an impressive range of business activities early in the twentieth century. Quarrying was a key feature of the business, supplying essential materials for the firm's activity. The company owned a quarry at Killycomain, which was acquired in the early 1900s and continued in operation until after the Second World War.[25] The proprietors initiated a wide array of ancillary enterprises designed primarily to serve their core business as building and civil engineering contractors. The firm developed a joinery department in Portadown, which provided a comprehensive range of timber products and remained an integral part of the business until the 1970s. The company also established its own brickworks at Seagoe, near Portadown, manufacturing clay bricks for its building or civil engineering projects.[26] The brickworks was an important asset to the firm throughout the first half of the twentieth century, although it declined after the Second World War and closed in the 1950s. Yet the company's business activity was not geared exclusively towards supporting its own construction operations: Collen also acted as builders' merchants, supplying building materials and hardware to other builders and private clients.[27] The builders' providers business, which was based in Portadown itself, later became a subsidiary company of Collen Brothers and remained a valuable element of the firm's portfolio until the 1960s.

Most of the additional strands of the business developed during this period were closely related to the core activity of the firm, namely building and civil engineering works. But there were occasional exceptions – at least one of the partners became involved in a maritime venture that was far removed from Collen's normal operations. Joseph Collen participated in a new merchant shipping company, investing in a venture to run a line of steamers between Newry and the ports of North America.[28] He was one of four directors of the company, which established in 1903 by a consortium of businessmen from Portadown, Newry and Rostrevor, County Down. Joseph became part-owner of a tramp steamer, the *Ulidia*, as a result of his venture into the world of merchant shipping.[29] The ultimate fate of this venture remains unclear due to a lack of documentary evidence, but it

---

25. Collen Papers Portadown, Speech by Joe Collen, 'My Job', p.3; Interview with Niall Collen, 5 February 2010.

26. Interview with Niall Collen, 5 February 2010.

27. Lyal Collen Papers, *The Collen Building Tradition*, p.2.

28. *Irish Times*, 'New line of steamers for Newry', 9 March 1903.

29. *Ibid.*; Lyal Collen Papers, *The Collen Building Tradition*, p.2.

was apparently a personal initiative on Joseph's part rather than a collective decision by the partners. His grandson Lyal Collen later described it as 'one small adventure' in the unfamiliar environment of merchant shipping.[30] But the venture remained an isolated interlude in the normal pattern of the company's activity. Collen Brothers was not tempted to explore new maritime horizons or to make collaborative ventures in merchant shipping a regular part of its activity.

The company was, however, ready to contemplate new initiatives that offered the prospect of expanding its core business. Collen Brothers branched out in a distinctive new direction in the first decade of the twentieth century, providing coastal lights for the first time. The firm concluded a contract with the Commissioners of Irish Lights on 18 October 1907 to provide the 'Sligo Harbour Lights'.[31] The project involved the construction of a new beacon in Sligo Bay, as an addition to the existing 12-foot high Metal Man lighthouse, which had guarded Sligo harbour since the 1820s. The works consisted of a beacon supported by a timber pile structure at Lower Rosses Point, along with small generator houses on Coney Island and Oyster Island.[32] The Sligo Harbour Lights was a small project, involving the extension of existing facilities for illuminating the approaches to the harbour. The company did build lighthouses later in the twentieth century, although Collen's association with lighthouse construction proved intermittent – its first venture in Sligo was the only contract obtained by the company with the Commissioners for Irish Lights under British rule and its next involvement in the area did not occur until the 1950s.

Much more significant was the company's commercial relationship with the British army, which was maintained and extended in the period before the First World War. Collen was operating in the Curragh military camp by 1899, when the company carried out the renovation and extension of the Wesleyan Soldiers' Home recently founded by the Methodist church.[33] The Methodist foundation was one of several similar institutions, which set out to provide for the physical and spiritual welfare of the soldiers in the Curragh: the Soldiers' Homes provided various comforts and recreational facilities for the men,

---

30. Lyal Collen Papers, *The Collen Building Tradition*, p.2.

31. River House Archive, Contract between the Commissioners of Irish Lights and Collen Brothers Ltd., *Works for Sligo Harbour Lights*, 18 October 1907, p.1.

32. *Ibid.*

33. *Irish Builder*, vol.41, *Tenders*, 15 November 1899, p.191.

Fire brigade station and tower, Curragh camp, completed by Collen Brothers in the early 1900s. Courtesy of the National Library of Ireland.

combined with a liberal allocation of prayer and hymns.[34] Collen benefited from the expansion of such quasi-religious institutions on military bases in the late nineteenth- and early twentieth-centuries. The company took charge in 1910 of the construction of a new Soldiers' Home, which was founded at the camp by Elsie Sandes, a Protestant lay missionary who enjoyed strong connections with the British military establishment.[35] The new institution, which was intended to serve as a centre of social recreation and spiritual support for the troops, opened its doors in October 1911.[36] Collen played a significant part in the rebuilding of the Curragh camp, which began before 1900 and continued through the first decade of the new century. The company undertook a series of projects at the Curragh in the early 1900s, building married quarters for soldiers in 1909, officers' messes for infantry and the Royal Engineers in the following year and new stables in 1914.[37] Collen also constructed a massive new water tower, which was completed in 1908 and became a characteristic

---

34. C. Costello, *A Most Delightful Station: The British Army on the Curragh of Kildare, Ireland, 1855–1922* (Cork, 1996), pp.123–4.

35. *Ibid.*; *The Building News and Engineering Journal*, vol.100, no.2928, 17 February 1911.

36. Costello, *A Most Delightful Station*, pp.124–5.

37. River House Archive, Sales Book for Collen Brothers 1900–49, pp.2–19.

feature of the Curragh's landscape.[38] A new fire station built sometime in the early 1900s formed another distinctive element of the firm's programme at the camp. Collen's association with the British army was by no means restricted to the Curragh. The firm constructed a new building for the Royal Hibernian Military School in the Phoenix Park between 1912 and 1914, and Collen was also involved in providing an officers' mess for the British garrison in Cork.[39] The company's highly productive business association with the British forces continued following the outbreak of the First World War in August 1914.

Water tower, Curragh camp, under construction, 1900. Lyal Collen Papers.

38. Costello, *A Most Delightful Station*, p.233; Lyal Collen Papers, *The Collen Building Tradition*, p.2.

39. River House Archive, Sales Book for Collen Brothers 1900–49, pp.9–14.

The European and ultimately global conflict which erupted in the autumn of 1914 generated new demands and opportunities for Collen Brothers. The company continued to operate in the Curragh during the First World War, completing building work on the rifle range between May and August 1915.[40] But Collen's primary contribution to the war effort was the manufacture of ammunition boxes ordered by the Ministry of Munitions, the British department established in 1915 to oversee the production of war materials and the mobilization of domestic industry in the service of a war economy. The company's venture into munitions production was a new departure, despite its long record as a building contractor for the British army. The Ministry of Munitions awarded several contracts for the production of military supplies to Irish firms and Collen acquired a share of the business.[41] The company fulfilled a series of orders for the British forces between the spring of 1916 and August 1918. Collen completed at least five major orders to provide ammunition boxes and cases to convey shells to the troops on the western front throughout the final two and a half years of the war.[42] The company also undertook building work for the Munitions Inspection Board near Kingsbridge in 1917.[43] While this last project was more in line with Collen's traditional profile as a building company, the manufacture of munitions boxes was undoubtedly a major part of the firm's business during the war. Although other companies based in Dublin, notably J.P. Good and T.R. Scott and Co, captured a much larger share of the contracts awarded by the Ministry of Munitions, it is apparent that Collen both contributed to and profited significantly from the British war effort.[44] The considerable stake acquired by the company in the munitions business undoubtedly helped to sustain Collen Brothers during the difficult economic conditions of the global conflict.

Yet the First World War also brought profound tragedy to the Collen family, as it did to many others in Ireland who participated in the wave of voluntary recruitment to the British army shortly after the outbreak of the war. Three members of the family perished in the course of the conflict. William Stewart Collen, a younger son of Joseph, worked in the family business before the war.

---

40. River House Archive, Sales Book for Collen Brothers 1900–49, p.17.

41. *Irish Times*, 'Munitions Work in Dublin: Charges Against Workmen', 18 April 1916.

42. River House Archive, Sales Book for Collen Brothers 1900–49, pp.21–8.

43. *Ibid.* p.22.

44. *Irish Times*, 'Munitions Work in Dublin: Charges Against Workmen', 18 April 1916; *Irish Times*, 'Dublin Munitions Tribunal: Forty-three strikers fined', 19 April 1916.

He volunteered for military service and secured a commission as a lieutenant in the Royal Inniskilling Fusiliers in September 1914. William Stewart was killed in August of the following year at the battle of Gallipoli aged only 26.[45] Two grandsons of John Collen also lost their lives in the Great War. Tom Shillington, the son of John's daughter Louisa and Major Graham Shillington, served as an officer in the Royal Irish Fusiliers; he died in August 1917 from wounds received in action on the western front.[46] The third member of the family to perish was John Collen, the namesake of the firm's founder, who was killed in 1918. The grim toll taken by the conflict on the Collen family reflected not only the fervent patriotism that motivated its younger members but also the family's staunch commitment to the crown and the institutions of the British state.

It was not the First World War but the ensuing conflict in Ireland that transformed dramatically the political and commercial context in which Collen Brothers operated. The general election in November 1918 brought the triumph of Sinn Féin and the eclipse of the Irish Parliamentary Party, the traditional representatives of constitutional nationalism, in most constituencies outside Ulster, where the Unionist Party maintained its position. As Sinn Féin had campaigned on a platform of republicanism and separation from the British state, its ascendancy within nationalist Ireland led rapidly to the establishment of the First Dáil in January 1919. The outbreak of armed conflict followed later in the same year between the Irish Volunteers who owed allegiance to the Dáil and the crown forces in Ireland. Collen sought to operate as normally as possible during the war of independence, continuing to fulfil contracts and serving traditional clients, including the British army at the Curragh. The firm maintained its operations in the Curragh throughout the conflict, building a memorial hall commemorating former soldiers and a generating station at the camp between 1920 and 1922.[47] Yet Collen mainly undertook contracts for civilian clients around this time, although whether this occurred by accident or design is not clear. The company initiated several new projects during the conflict, including a new box-making factory in the South Lotts area of Dublin's inner city in 1920: the new building was initially constructed as a timber structure on concrete foundations, due to a severe shortage of building materials.[48] Collen was also engaged in building work for

---

45. *Irish Times*, 'Lieutenant W.S. Collen', 19 August 1915.

46. *Irish Times*, 'Army list', 21 August 1917.

47. River House Archive, Sales Book for Collen Brothers 1900–49, p.34.

48. *Ibid.* pp.27–32; *Irish Builder*, vol.62, 'Building News', 10 April 1920, p.229.

a church at Dundrum for the Church of Ireland and undertook works at the Ulster Bank in Ferbane around the same time, underlining the considerable breadth of operations maintained by the firm in the midst of the conflict.

The firm's venture into the production of munitions boxes ended with the First World War and there is no record of Collen fulfilling orders for any military supplies during the war of independence. While John Collen remained chairman of Collen Brothers until his death in 1921, Joseph was primarily responsible for managing the business through the successive crises of the decade.

The eventual outcome of the conflict between the Irish nationalist move-ment and the British state presented Collen with the most formidable challenges since its foundation. The Anglo-Irish Treaty in December 1921 provided for a self-governing Irish state for the first time but also confirmed the division of the island. The Irish Free State came into being in 1922, while the devolved Parlia-ment in Stormont, dominated from the outset by the Unionist Party, exercised authority over the six counties of Northern Ireland. Collen faced an apparently daunting prospect in adapting to the new political dispensation. The company operated on an all-Ireland basis and had completed an extraordinary variety of projects in all four provinces over the previous half-century. Moreover, Collen Brothers was a Protestant family business, which had enjoyed profitable connec-tions with the British administration in Ireland and elements of the traditional Unionist aristocracy: those sources of patronage were abruptly removed, at least south of the border. The nationalist struggle for independence had swept away the traditional political and social environment within which Collen had pros-pered. The company was obliged to operate in two distinct and mutually antag-onistic political jurisdictions on different sides of the newly established border. Perhaps more significantly, the directors had to cope with the disappearance of many of the familiar landmarks they had known for their entire working lives – the union with Britain, the presence of the British army at the Curragh, and, to a lesser extent, the patronage of prominent aristocratic figures.

Yet Collen succeeded in adapting to the new dispensation with impres-sive speed and dexterity. The partition of Ireland was followed by a changing of the guard at the top level of the company, although this process occurred gradually and was not fully completed until the late 1920s. John Collen, who was still the chairman of the board, died in May 1921 at the age of 84; his death symbolized the end of an era for Collen Brothers.[49] His younger brother

---

49. *Irish Builder*, vol.63, *Obituary: Mr. John Collen, D.L., Portadown*, 4 June 1921, p.390; *Irish Times*, 'Mr. John Collen D.L', 21 May 1921.

Sale of Killycomain House, 'Gentleman's Residence', Portadown, 25 June 1924; the house was sold three years after the death of John Collen. Courtesy of Niall Collen.

Joseph succeeded him as chairman of the firm, but gradually withdrew from active involvement in managing the business over the following decade. He resigned as a director in 1930, although he remained nominally at the head of the firm until 1941.[50] Richard held the post of joint managing director and remained active in the business until his death in February 1927, but the de facto management of the company passed to younger members of the family in the 1920s.[51] David Collen, the youngest of the founding generation, remained a director of the company, but was not greatly involved in

---

50. Collen Papers Portadown, *Case for Counsel to advise Mr. J.B. Collen*, 1943.

51. *Irish Times*, 'Mr. Richard Collen', 8 February 1927.

the daily management of the business during this period.[52] Instead it was Joseph Collen's two surviving sons, John Black and Joseph Harcourt, who emerged as the leading figures of the new generation. Both men had already been deeply involved in the company for over two decades, acting as junior managers on the Portrane asylum project.[53] Indeed, Harky lived in a house at the Curragh shortly before the First World War, supervising the company's extensive building projects at the camp.[54] Following a lengthy apprenticeship, the two men took over the running of the firm in the 1920s. John Black (who was widely known as Jack) took charge of the business in Portadown, while his younger brother Harky became the effective head of the company in Dublin.[55] The two brothers played the central role in managing the business on different sides of the border during the inter-war period. Collen Brothers remained formally a single company, but a pragmatic division of labour was adopted between the two leading figures – each director enjoyed a considerable degree of autonomy in managing his side of the business, although they also maintained regular contact with each other.

The correspondence in the Collen papers gives a flavour of the practical difficulties created by the newly established border and illustrates the efforts made by the Collen family to surmount such obstacles through mutual collaboration. The new political settlement affected the most mundane details of life as well as more significant business decisions. Harky Collen was obliged to seek the permission of the new Stormont government, led by Sir James Craig, to bring his car into Northern Ireland on a regular basis – significantly, he had little difficulty in securing the agreement of the government to facilitate his business travel. Harky wrote to the Surveyor of Customs in Newry on 28 February 1924 to apply for a pass for 'the temporary importation of my Motor Car … into Northern Ireland as per arrangement in existence with Northern Ireland Government.'[56] His surety, who signed a bond on his behalf, was his brother John Black Collen, 'Contractor, Hanover St., Portadown'.[57] The mundane nature of the request does not diminish the importance of the formal

52. Lyal Collen Papers, Lyal Collen Note, *Management*, p.1.

53. Lyal Collen Papers, Lyal Collen Note, *Management*, p.2.

54. NAI, Census of Ireland 1911, *Household Return for Harcourt Collen*.

55. River House Archive, Letter books for Collen Brothers, 1919–27.

56. River House Archive, Letter book for Collen Brothers 1923–24, J.H. Collen to Surveyor of Customs, Newry, 28 February 1924.

57. *Ibid.*

arrangement made by Collen Brothers with the Northern Ireland government to facilitate their business activity. The fact that such an arrangement was made in the first place underlined the expectation that regular travel across the border on business by the directors would be required. It was equally apparent that the two brothers sought to maintain an underlying collaboration within the family business despite the political obstacles. Overall, the Collen family moved swiftly to adapt to the reality of a newly partitioned island, not least by making arrangements for the dual administration of the family business.

The gradual transition to a new generation undoubtedly facilitated Collen's adaptation to new political circumstances, but the company also benefited from external factors, notably the compelling demands for reconstruction and social progress after the First World War and the military conflict in Ireland. The need for post-war reconstruction was acutely felt throughout Europe in the 1920s, although it took a somewhat different form in Ireland, where it meant recovery from the ravages of the war of independence and civil war more than the previous global conflict. Reconstruction in Dublin often meant the literal rebuilding of commercial property and public buildings destroyed during the successive phases of military conflict. The city centre of Dublin had been devastated twice, first by the British forces in their campaign to crush the 1916 Rising and then by the struggle between the Free State army and the anti-Treaty IRA in 1922. Collen secured a considerable volume of work in the war-torn centre of the city during the 1920s. The company constructed a new building for the Ulster Bank in O'Connell St, which had been devastated during the 1916 Rising. The completion of the new premises in 1923 marked the final step in the reconstruction of Lower O'Connell Street.[58] Collen was also involved in the restoration of the Four Courts, which had been badly damaged by the artillery of the Free State army in 1922: the company rebuilt the Land Registry at the Four Courts in the late 1920s.[59] The economic and administrative imperatives to restore the centre of the city created a ready market for building contractors in the first decade of the Irish Free State.

Perhaps equally important to Collen was the wider social demand for public housing in the 1920s. The company had previously engaged in the building of country houses and aristocratic homes, as well as more modest

---

58. River House Archive, Sales Book for Collen Brothers 1900–49, p.33; *Irish Builder*, vol.65, 'Building News', 10 February 1923, p.101.

59. River House Archive, Sales Book for Collen Brothers 1900–49, p.41; *Irish Times*, 'Dublin Circuit Court', 9 November 1929.

dwellings for private clients, but it developed a significant involvement in public housing projects for the first time during the 1920s. The firm issued a tender to undertake a public housing scheme at Donnelly's Orchard, which was initiated by Dublin Corporation in 1923.[60] Collen lost out on the contract to build eighty-four new council houses to their competitors, H. and J. Martin, but won the tender to provide the ancillary roads, paths and drains on the same site.[61] Collen also built cottages for veterans of the First World War during the early 1920s in south Dublin, Kildare and Naas, which were financed initially by the British government.[62]

But the most significant housing projects provided by the company around this time owed more to non-profit housing initiatives by Church of Ireland clergy than any public scheme. Rev. David Hall, a Church of Ireland rector in East Wall and housing reformer, founded the St Barnabas Public Utility Society in January 1920 as 'a Christian endeavour to meet and correct some of the appalling housing conditions of Dublin.'[63] Hall's initiative created the first operational public utility society in Ireland. Its objective was to build housing for the working class who were living in frequently atrocious conditions in the Dublin slums and allow them to purchase their rented homes over a twenty year period. Collen built sixty-two houses for the St Barnabas Society on West Road in East Wall between 1924 and 1925.[64] The scheme was designed to accommodate 'better class workmen' and the houses were of a higher standard than the Corporation developments of the time, providing hot water, baths and a garden.[65] Soon afterwards, the company took responsibility for delivering a similar project on the site of the old Linen Hall. The Linenhall Public Utility Society was founded by Rev. E.J. Young and Rev. R.S. Griffin, with strong encouragement from Hall, at the beginning of 1926.[66] Collen secured the contract with the Society to build sixty-three houses on the

60. *Irish Times*, 'Dublin Corporation Dwellings', 3 May 1923.

61. River House Archive, Sales Book for Collen Brothers 1900–49, p.122.

62. *Irish Times*, 'Houses for Irish Ex-Servicemen', 29 December 1921.

63. Ruth McManus, 'The "Building Parson" – The role of Reverend David Hall in the solution of Ireland's early twentieth century housing problems', *Irish Geography*, vol.32(2), 1999, pp.87–98.

64. River House Archive, Sales Book for Collen Brothers 1900–49, p.36.

65. Ruth McManus, 'The "Building Parson" – The role of Reverend David Hall in the solution of Ireland's early twentieth century housing problems', *Irish Geography*, vol.32(2), 1999, p.95.

66. *Irish Times*, 'House Ownership in Dublin: Linen Hall building scheme', 15 October 1926.

Linen Hall site at a cost of approximately €31,000.[67] The project, which was completed in 1927, was funded in part by private subscriptions raised by the Society, supplemented by fixed-sum grants from Dublin Corporation under legislation recently passed by the Dáil. In accordance with the Society's plans, the company provided four- or five-room houses, which were considerably more spacious and comfortable than the overcrowded and substandard tenements that passed for accommodation in much of the city during the 1920s.[68] The two projects completed by Collen were relatively small, but testified to an astute decision by the company to collaborate with the public utility societies, which formed part of a rapidly expanding movement in the 1920s. The movement initiated by Rev. Hall made a significant contribution to increasing Ireland's housing stock in the interwar period.[69] Collen's association with the public utility societies was a case of enlightened self-interest, as the emergence of the societies offered an important business opportunity. The schemes provided valuable contracts for the company, which were subsidized by the new Irish state; the building of the new houses in East Wall could also benefit Collen by providing decent accommodation for skilled workers, including actual or potential employees of the firm. Collen benefited from its collaboration with public and private efforts to remedy chronic housing shortages and improve appalling housing conditions under the new Irish Free State.

The foundation of the new state also gave Collen Brothers an unexpected opportunity to re-establish its association with the RDS, which had been in abeyance since the early 1900s, when the Society employed other contractors for the building of the Art Industries Hall at Ballsbridge. The new Provisional Government took over Leinster House in 1922, forcing the Society to abandon its headquarters in the city centre and relocate fully to Ballsbridge.[70] The RDS was obliged to initiate a major rebuilding programme at Ballsbridge, supported by a grant of £68,000, which they received from the Provisional Government. Collen was an unintended beneficiary of the new government's action and the Society's enforced relocation. Lucius O'Callaghan of O'Callaghan and Webb architects designed the new buildings at Ballsbridge, while Collen acted as

---

67. *Irish Times*, 'The Church of Ireland', 20 April 1926.

68. *Irish Times*, 'House Ownership in Dublin: Linen Hall building scheme', 15 October 1926.

69. Ruth McManus, 'The "Building Parson" – The role of Reverend David Hall in the solution of Ireland's early twentieth century housing problems', *Irish Geography*, vol.32(2), 1999, p.96.

70. *The Royal Dublin Society 1731–1981*, ed. Desmond Clarke and James Meenann (Dublin, 1981), p.73.

The new frontage of the Royal Dublin Society, 1928. Lyal Collen Papers.

the general building contractors – the company completed a series of major projects for the reconstruction and extension of the Ballsbridge site between 1924 and 1928. The process of redevelopment began in 1925 with the reconstruction of the Art Industries Hall, which was converted into a concert and recital hall, providing seating for over 1200 people.[71] The enlarged hall was laid out in aisles marked by columns in the Greek Ionic style, and the side aisles were lined with bookshelves, which accommodated part of the Society's library. Collen also converted the corridor leading to the original Art Industries building into a large hall, where the art industries exhibits could be displayed.[72] The transformation of the Art Industries Hall was only the first stage of a wholesale reconstruction of the RDS in which Collen played an integral part.

The most extensive phase of the rebuilding programme began in 1926, when Collen was given responsibility for providing a new frontage to the RDS, constructing new administration buildings and linking the new structures together. The previous layout consisted of a series of showgrounds set well back from the road, with a main central entrance; the new buildings were provided in front of the existing show halls on either side of the main

---

71. *Irish Times*, 'Dublin Spring Show: Increased Entry and New Features', 30 April 1925; *Irish Times*, 'Royal Dublin Society: Many Improvements at Ballsbridge', 24 January 1925; *Irish Times*, 'New Buildings at Ballsbridge Show Grounds', 30 July 1925.

72. *Irish Builder*, vol.67, 'Topical Touches', 8 August 1925, p.633.

entrance.[73] This meant the demolition of the original red-brick frontage of the hall constructed in 1880. Instead Collen, acting on the basis of O'Callaghan's designs, constructed new granite-faced buildings of a neo-Georgian style flanking the main entrance, which was faced up to bring it into harmony with the surrounding structures.[74] The work demanded a very high finish in the details of the stonework and joinery: the granite stonework was bedded in a lime mortar with very fine limits so that the setting out of each stone involved a high degree of care and precision.[75] The project entailed the reconstruction of the central features of the frontage, with the redeveloped frontage of 562 feet now even longer than before.[76] Collen provided a new west wing, which incorporated the main library and science laboratories, as well as reading, writing and smoking rooms for members of the Society.[77] The east wing was rebuilt to provide accommodation for the administrative headquarters of the Society, including offices, a committee room and the council chamber.[78] The frontage and adjoining buildings were completed by the summer of 1928; the achievement of the architect and builders was widely acclaimed both by national newspapers and trade journals. The *Irish Times* commented on 2 August 1928 that the new buildings 'form a charming feature and a welcome addition to the public buildings of note in Dublin.'[79] *The Architects' Journal*, taking somewhat longer to be won over by the architectural charms of the new RDS, praised the 'modern and individual character' of the new frontage in June 1931.[80]

Perhaps more significantly for Collen, the council of the RDS was satisfied enough with the performance of its building contractors that another major contract was awarded to the company well before the frontage became a reality. The council decided in December 1926 to build a new stand in the jumping arena at Ballsbridge and appointed Collen as the main contractor for the project: the work began even before the formal decision by the council,

73. *The Architects' Journal*, vol.73, 17 June 1931, pp.849–50.

74. *Ibid.*

75. Lyal Collen Papers, *Royal Dublin Society*, p.1.

76. *Irish Builder*, vol.68, 'Topical Touches', 9 January 1926, p.5, *The Architects' Journal*, vol.73, 17 June 1931, pp.849–50.

77. *Ibid.*

78. *The Architects' Journal*, vol.73, 17 June 1931, pp.849–50; *Irish Times*, 'Improvements at Ballsbridge', 20 April 1927.

79. *Irish Times*, 'Building and Reconstruction: Headquarters of the Royal Dublin Society', 2 August 1928.

80. *The Architects' Journal*, vol.73, 17 June 1931, pp.849–50.

as the Society's committee for agriculture gave the company the green light to proceed earlier in the year.[81] The new stand was a ferro-concrete structure with a steel frame, which was built over the existing twelve-step terrace in the enclosure. Collen carried out the concrete work, while the steel frame was provided by A. and J. Main Ltd.[82] The Anglesea Stand offered covered accommodation for over 2000 spectators when it was opened in July 1927.[83] Yet the demand for seating was so great at the Horse Show in 1927 that within weeks of opening the new stand the council agreed to extend it, with the aim of doubling its capacity.[84] Collen extended the Anglesea Stand by a further 300 feet so that it ran the whole length of the jumping arena – the newly completed stand was the largest reinforced concrete structure in the Ireland of the 1920s.[85] The company played a central part in the vast rebuilding programme undertaken by the Society in the 1920s and made a crucial contribution to the development of the modern RDS, which largely took shape in this period.

The success of Collen Brothers in adapting to the new political dispensation south of the border was vividly highlighted in 1932 when the company resumed work at the Curragh – this time for the Irish army. Collen returned to the Curragh barely a decade after the withdrawal of its former client, the British forces, from the camp. The firm won a contract with the Commissioners of Public Works in July 1932 to conduct large-scale alterations to Pearse barracks.[86] The scheme was designed to adapt the barracks for the establishment of a new military college at the Curragh.[87] Collen completed the project in 1933 at a price of over £7269, which was broadly similar to the value of other important contracts that the company had undertaken at the camp before 1922.[88] Yet more significant than the financial value of the contract was the reappearance of Collen in military building projects at the Curragh; the company evidently did not suffer by its previous association with the British army.

81. Lyal Collen Papers, Kevin Bright to Lyal Collen, 6 May 1991; *Irish Builder*, vol.68, 'Building News', 27 November 1926, p.891.

82. *Irish Times*, 'Improvements at Ballsbridge', 20 April 1927.

83. *Irish Times*, 'The Coming Horse Show at Ballsbridge', 9 July 1927.

84. *Ibid*.

85. Lyal Collen Papers, *Royal Dublin Society*, p.1.

86. *Irish Times*, 'Government Contracts Placed', 21 July 1932.

87. IAA, PKS 0185 A09, 77/1/A9, Patterson and Kempster Surveyors, *Contract with the Commissioners of Public Works, Ireland, for Pearse Barracks, Curragh Camp*, April 1931.

88. River House Archive, Sales Book for Collen Brothers 1900–49, p.42.

New Members Hall at the RDS, completed between 1926 and 1928.
Courtesy of the Library, Trinity College, Dublin.

It is apparent that Collen Brothers adapted to the political division of the island with remarkable success, not least by establishing a new management structure giving considerable autonomy to different branches of the company. Despite the firmly Unionist orientation of its leading members in the previous generation, no lingering political sentiment was allowed to stand in the way of the survival of the business. Collen took full advantage of commercial opportunities in the Irish Free State, rekindling its association with traditional clients such as the Royal Dublin Society, and acquiring new ones, including the public utility societies and the Board of Public Works of the new state. The company undoubtedly benefited from a variety of social and economic demands in post-independence Ireland, notably the necessity for reconstruction after the ravages of successive military conflicts and increased public and private concern to improve the appalling housing conditions for the working class in Dublin. But the firm's success in securing such a substantial volume of business in relatively difficult political and economic conditions highlighted the resilience and pragmatism of Collen Brothers. It appeared that the leading directors had succeeded in maintaining the fortunes of the company, while preserving the family business as an all-Ireland entity in the face of war, sectarian strife and partition. Yet the border had enforced a practical division of responsibility within the company on political and geographical lines, which foreshadowed the later formal separation between the northern and southern branches of the business.

THREE

Amicable Separation

It was apparent by the late 1920s that the political reality of the independent Irish state was far less threatening to Collen Brothers than the prevailing economic conditions. The world economic depression impinged on the firm's activity on both sides of the border. Most large-scale projects undertaken by Collen Brothers in the Irish state during the 1930s were initiated by public authorities and non-profit housing trusts, although the firm also benefited from repeat work with traditional private clients such as the RDS. The company's activity in Northern Ireland followed a broadly comparable pattern at the height of the economic depression, but the nature of the work undertaken by the two sides of the business diverged significantly during the Second World War. The practical division of responsibility between north and south was the dominant feature of the firm throughout this period. This appeared to work reasonably well in practice, but internal debates about the future of the company were not fully resolved and ultimately the division of the firm itself was postponed for a generation rather than avoided.

The pragmatic division of activity worked largely because of the close collaboration between the two brothers who were also the most active and influential directors, Harky and Jack Collen. Harky held the post of managing director for much of this period, but in practice he managed the business in

Jack (left) and Harky Collen, pictured in front of the offices of Collen Brothers, Hanover St., Portadown. Courtesy of Niall Collen.

Dublin, while Jack Collen fulfilled the same function in Portadown. Yet while the daily management of the business was divided between the two men, Collen Brothers remained a single firm governed by the same board of directors. The registered office of the company remained in Hanover Street, Portadown, which was still formally the headquarters of the business. All of its annual general meetings were held in Northern Ireland until the late 1940s, with the exception of a single meeting in 1943, which took place in Dublin; the firm's long-serving secretary, W.T. Church, was also based in Portadown.[1] Moreover, Harky and Jack Collen shared a sense that it was still a single business, collaborating together and involving their own children in the firm. Just as the two men had taken over the direction of the company from the founding partners, so younger members of the next generation began to make their presence felt during this period. Jack Collen, who married Eleanor Bassett from Cork, had two sons, Anthony (Tony) and Joseph Bassett (Joe).[2] Both men became involved in Collen Brothers at an early age and later took charge of

1. Collen Papers Portadown, *Case for Counsel to advise J.B. Collen*, 1943, pp.1–4.
2. *Ibid.*; see Collen family tree.

the northern branch of the firm. Harky Collen and his wife Mary Arnott Neill raised four children, including three sons, Desmond, Standish and Lyal, and one daughter, Dorothy.[3] All four would be shareholders of Collen Brothers at various times, but it was Standish and Lyal who emerged as the central figures within the Dublin branch of the firm following the Second World War. While its practices and management structures might change, it was apparent that Collen Brothers remained first and foremost a family business.

SURVIVING THE 1930S

The company faced a generally unfavourable economic outlook in the Irish state even before the world economic depression that was triggered by the Wall Street crash in 1929. The *Irish Builder* drew attention to the shortage of building materials in the Irish market as early as 1920.[4] The building trade also suffered from an inadequate supply of skilled workers and the inability of many building contractors to raise sufficient capital to undertake substantial projects. A report by Gerald Sherlock, the Dublin city manager, in April 1934 identified the major obstacles to the rapid completion of urban housing schemes in the state as 'the limited capacity of the building trade in respect of the shortage of skilled labour, as well as the shortage of materials (mostly Irish manufacture) and the problem involved in raising large capital sums.'[5] The city manager complained that 'great difficulties had been experienced through shortage of materials, such as bricks, tiles, slates, concrete pipes, rain-water goods, ranges, timber etc.'[6] While these problems were undoubtedly exacerbated by the Great Depression, most were also long-term issues that predated the economic crash of 1929. The scarcity of essential building materials can be attributed in part to the shortcomings of Irish manufacturing industry during the interwar period. The high level of emigration to Britain or the United States, which remained a persistent feature of Irish society in the first generation of the independent state, undoubtedly contributed to the difficulties facing building contractors in securing a sufficient supply of skilled workers. The shortage of private capital to support major projects was an international

---

3. Lyal Collen Papers, Note on Collen family, 1950, p.7; see Collen family tree.

4. *Irish Builder*, vol.62, 'Building News', 10 April 1920, p.229.

5. *Irish Times*, 'The Corporation's Plans', 26 April 1934.

6. *Ibid.*

phenomenon by the 1930s, but it was also a consistent problem in restricting the economic development of the Irish state up to the 1960s. The bulk of the activity in the construction business in this period was driven by the state and local government, which acted to develop public works, including large-scale housing projects and other forms of social infrastructure, notably the building of hospitals.

Collen acquired a substantial share of the construction business created by the state's programme of hospital building and reconstruction in the 1930s. The election of the first Fianna Fáil government, led by Eamon de Valera, in 1932 brought a significant increase in spending on social services, including health.[7] The new government initiated an impressive programme of hospital building, financed largely by the proceeds of the Irish Hospitals Sweepstake. The financial success of the sweepstakes and the establishment of the Hospitals Trust Fund under the authority of the Minister for Local Government in 1933 to co-ordinate the distribution of sweepstake funds ensured that considerable resources were available for the development of local authority hospitals.[8] Collen had built hospitals and sanatoria before the First World War, but its involvement in hospital building then became dormant until the early 1930s. Collen's first major venture in hospital building for over two decades was the reconstruction of the fever hospital at Abbeyleix.[9] The original rationale behind the fever hospitals was the isolation of patients from society to prevent the wider spread of infectious diseases, but it was apparent by the 1930s that such isolation hospitals had achieved little or no success in preventing the spread of contagious diseases, while the older fever hospitals had also facilitated the spread of illness due to cross-infection among their patients.[10] The medical establishment and local authorities sought to upgrade and modernize fever hospitals in this period, while widening the range of diseases treated in such hospitals and maintaining a strong element of isolation for seriously ill patients. Collen benefited from a sustained attempt by the government and local authorities to replace unsuitable or dilapidated hospital buildings, which often dated back to the previous century.

---

7. Diarmaid Ferriter, *The Transformation of Ireland: 1900–2000* (London, 2004), pp.398–9.

8. Marie Coleman, *The Irish Sweep: A History of the Irish Hospitals Sweepstake 1930–87* (Dublin, 2009), pp.61–2; *Irish Times*, 'Leix County Hospital: Minister's advice to public bodies', 18 December 1936.

9. *Irish Builder*, vol.75, 'Contracts', 7 October 1933, p.870.

10. *Irish Times*, 'Modern planning and design: opening ceremony takes place today', 21 July 1938.

The company encountered some local resistance to its tender for a new hospital building at Abbeyleix. When the Board of Health for Co. Laois considered the tenders in November 1933, its members were divided, with several favouring the acceptance of a tender from a local contractor, John Fitzpatrick of Portlaoise.[11] The local contractor submitted the lowest tender, but his specifications for the job proved unacceptable to the Department of Local Government and Public Health, either because Fitzpatrick cut some corners in his tender to reduce the price or due to a suspicion that local patronage would not produce the best outcome to the contest. Following a debate over the tender at the Board's meeting, the members decided by majority vote to accept Fitzpatrick's tender, subject to the sanction of the Department of Local Government, but added the crucial stipulation that 'in the event of the Local Government Department declining to sanction Mr Fitzpatrick's tender, the next lowest tender, Messrs. Collen Brothers, be accepted, subject to sanction'.[12] The tenor of the debate indicates that the department's objections to the local contractor's tender were already anticipated – in any event, the decision virtually guaranteed that Collen would receive the contract. It was a large-scale project involving the demolition of the existing building and the construction of a new fever hospital with additional accommodation for patients, along with a mortuary, gate lodge, water tower and boundary walls.[13] The main buildings were completed over a two-year period and the new hospital was opened on 17 September 1936.[14] The Board of Health in Co. Laois emerged as an important client for Collen in the 1930s. The company acquired a similar contract in the summer of 1936 to provide a new county hospital at Portlaoise designed by Scott and Good architects.[15] This proved a more protracted undertaking and the work was completed only in 1941, although the hospital itself was in operation by 1938.[16]

11. *Irish Times*, 'Contracts for New Fever Hospital', 10 November 1933.

12. *Ibid.*

13. *Irish Builder*, vol.75, 'Contracts', 7 October 1933, p.870.

14. River House Archive, Note by Henry Overend, Sales Book for Collen Brothers 1900–49, p.48.

15. *Irish Times*, 'Contracts placed for building hospitals, school and hundreds of houses', 11 June 1936.

16. IAA, Scott, Tallon and Walker Drawings Collection, vol.1, 79/10.5/1–71, Messrs. Scott & Good, *Designs for Portlaoise County Hospital, Co. Laois, 1933–41*; River House Archive, Sales Book for Collen Brothers 1900–49, p.52.

Fever Hospital, Naas, Co. Kildare, 1939. Collen Brothers photo album.

Collen's participation in hospital building received favourable public attention as a result of the firm's involvement in the replacement of the fever hospital at Naas. The medical advisers of the Department of Local Government had condemned the existing fever hospital in the town as unfit for purpose.[17] Kildare County Council was obliged to seek tenders from 'competent builders' to replace the existing structures and Collen was awarded the contract in May 1935 to build 'a new hospital for communicable diseases' at Naas.[18] The new hospital in Naas was intended to minimize risks of infection, by providing for segregation of patients suffering from different diseases, adequate bed spacing within wards and increased numbers of medical staff. The design of the hospital, undertaken by Vincent Kelly, an architect from Dublin, incorporated separate accommodation and facilities for patients with different illnesses, along with several isolation rooms to facilitate the observation of patients where a diagnosis of their condition was unclear.[19] The new hospital also provided for a significant increase in capacity, accommodating thirty-eight patients as opposed to a maximum of twenty-three in the old building. Collen completed the project on schedule by the summer of

17. IAA, Accession 2009/100, Album of press cuttings relating to Vincent Kelly; *Irish Press* 'Kildare's New Hospital', 21 July 1938.

18. IAA, Accession 2009/100, Album of press cuttings relating to Vincent Kelly; *Leinster Leader*, 'Notice to Builders', 18 April 1935; River House Archive, List of tenders, *New Fever Hospital – Naas*, 22 May 1935.

19. *Irish Times*, 'Modern planning and design: opening ceremony takes place today', 21 July 1938.

1938 and Seán T. O'Kelly, the Tánaiste and Minister for Local Government, formally opened the new hospital on 21 July 1938.[20] The opening ceremony generated highly positive publicity for Collen, as its efforts attracted favourable notice in two national newspapers. The reports on the opening of the new hospital by the *Irish Times* and the *Irish Press* commented on Collen's role in virtually identical terms, taking their cue from statements made by O'Kelly and the architect: 'The general contractors … were Messrs. Collen Bros., East Wall, Dublin, who carried out the work in record time and with great satisfaction to all concerned.'[21] While Collen had no role in planning the project, the company was credited by most observers with implementing the architect's design in a highly professional fashion. The company contributed significantly to the programme of hospital building in this period and was also a substantial, if indirect, beneficiary of the funding allocated to local authority hospitals from the sweepstakes. All three of the new hospitals built by Collen in the 1930s were financed from sweepstake funds.[22] Hospital building emerged as the most important single element of Collen's activity in the independent Irish state during the 1930s.

The construction of urban housing schemes remained a significant feature of the firm's work throughout this period, although it had relatively little involvement in providing local authority housing in Dublin. This was not due to lack of willingness on the company's part: Dublin Corporation promoted several major public housing schemes in the 1930s, but Collen was not successful in tendering for these projects. The company tendered for the extension of the council housing development at Donnelly's Orchard in 1933, but failed to secure the contract, which went instead to H. and J. Martin.[23] But towards the end of the decade Collen undertook a substantial housing project for the Iveagh Trust, a charitable foundation originally established in 1890 by Sir E. C. Guinness, the first Earl of Iveagh. The Trust set out to provide housing for low-income families at affordable rents and with decent

---

20. IAA, Acc. 2009/100, Album of press cuttings relating to Vincent Kelly; *Irish Press*, 'Kildare's New Hospital' 21 July 1938; *Irish Times*, 'New Kildare Hospital: Formal opening by Minister', 22 July 1938.

21. *Ibid.*; *Irish Press*, 'Kildare's New Hospital' 21 July 1938; *Irish Times*, 'Modern planning and design: opening ceremony takes place today', 21 July 1938.

22. Coleman, *The Irish Sweep*, pp.78–9.

23. *Irish Times*, 'Housing Loan and Tenders', 4 February 1933; River House Archive, *Bill Of Quantities for the Erection of 148 Houses at Donnelly's Orchard Extension Area*, Corporation of Dublin, October 1932.

amenities.[24] Collen built four new blocks of flats for the Trust, close to its original development for artisans' dwellings on Kevin Street, between 1938 and 1941.[25] The first block, which was ready for occupation in February 1940, consisted of fifty-seven flats, many of them equipped with separate bathrooms and kitchens, as well as a number of shops. The remaining three blocks, which were completed in the early 1940s, provided a further sixty apartments.[26] Overall the new housing scheme accommodated about a thousand people: it also incorporated shops, gardens and a playground for children between the flats.[27] The project reflected the progressive philosophy of the Trust, providing greater amenities and recreational facilities than most housing schemes of the time. The Iveagh Trust development was the most significant housing project completed by Collen in this period.

While the company's work during this time was generally dominated by major building contracts involving the provision of social infrastructure, Collen also undertook a variety of smaller projects for private clients. The firm became strongly associated with the delivery of facilities for racecourses in the 1930s. Collen built a wooden stand and a new building for the Tote at Fairyhouse racecourse in 1931.[28] The company also provided wooden Tote facilities for race meetings throughout the country, including traditional venues such as the Curragh, Punchestown and the Phoenix Park. Yet Collen's activity extended far beyond major race venues: the company provided Tote buildings at Mullingar, Mallow, Ballinasloe, Baldoyle and Gowran Park in the course of a single month in June 1930.[29] Such a volume of business was in no way exceptional and Collen remained deeply engaged in the building of racecourse facilities throughout the following generation. The Second World War brought no cessation in popular enthusiasm for race meetings: the company completed Tote facilities in Limerick city, Thurles and Tramore in 1945, as well as securing repeat business at Fairyhouse and providing new stables at Baldoyle.[30] The breadth of Collen's activity was striking, as the firm was undertaking building

24. *Irish Times*, 'The Iveagh Trust Flats', 12 January 1940.

25. River House Archive, Sales Book for Collen Brothers 1900–49, pp.54–5; *Irish Times*, 'Iveagh Trust Enterprise', 12 January 1940.

26. *Irish Times*, 'Work of Iveagh Trust', 26 February 1940.

27. *Irish Times*, 'Work of the Iveagh Trust: Play Centres, Homes, Meals and Recreations', 19 March 1941.

28. River House Archive, Sales Book for Collen Brothers 1900–49, p.46.

29. *Ibid.* pp.44–5.

30. *Ibid.* pp.63–8.

work at racecourses in most counties around this time.[31] The profits involved were usually relatively small, but the volume of the business and the low overheads involved in the manufacture of Tote buildings made it a worthwhile commercial venture. The directors clearly identified the pervasive nature of horse racing in Irish social life as a valuable seam to be tapped.

Stand and Tote building for Fairyhouse racecourse, pictured in 1947.
The stand was designed and built by Collen within eight weeks.
Collen Brothers photo album.

It certainly did no harm to the company's prospects that Harky Collen was an enthusiastic patron of the track and himself owned racehorses at various times.[32] Harky was indeed a successful participant in point-to-point races for over thirty years, as well as an active member of the Ward Union Hunt. He was also a steward of the RDS, where he served as a member of the organizing committee for the Horse Show in the 1930s.[33] It was hardly surprising that

31. *Ibid.* pp.44–5.

32. *Irish Times*, 'Jumping at the Show', 4 August 1932; *Irish Times*, 'Ballymacad Point-to-Point Races', 26 March 1949.

33. *Irish Times*, 'J.H (Harky) Collen: An Appreciation', 29 March 1963; *Irish Times*, 'Royal Dublin Society: Meeting of Horse Show committee', 6 August 1931.

Collen Brothers maintained its long-term commercial association with the RDS, which reflected the family's close connections with the racing fraternity, as well as the firm's enduring contribution to the development of the Society's grounds over the previous half-century. The company took responsibility for the expansion of the Pembroke Hall, which formed part of the Ballsbridge show grounds, in 1938: the width of the hall was doubled by the spring of 1939 to provide additional space for the major showpiece events organized by the Society.[34] Collen Brothers in Dublin weathered the economic storms of the depression by taking full advantage of new commercial opportunities, offered especially by the fortuitous expansion of capital investment in hospital construction, but also by cultivating traditional clients and well-established business or family connections.

The northern branch of the company displayed similar resilience during this period. Collen extended its commercial tentacles throughout Northern Ireland in the 1920s, providing county council housing, sewerage and water-works schemes.[35] The company also won contracts for the construction of post offices from the Stormont administration, building a new post office in Strabane, Co. Tyrone, in 1928.[36] Yet the construction industry in Northern Ireland was severely affected by the depression and the firm struggled to find much new work during the 1930s. Jack's son Joe Collen, who entered the firm in the early 1930s, having completed a B.Sc in civil engineering at Queen's University, Belfast, remembered it as 'a very lean time, there was very little work on'.[37] The first project in which Joe was involved was the installation of a watermains scheme at Magherafelt in Co. Derry. He acted as the engineer for the job, which proved a difficult undertaking, not least because it occurred in the era before the use of heavy machinery and was completed entirely by workers using shovels.[38]

The firm north of the border was sustained by state contracts and repeat business, often from public authorities but also from traditional private clients. The company maintained a central role in providing public works for the local authorities in north Armagh. Collen Brothers installed a new

---

34. *Ibid.* p.53; *Irish Times*, 'RDS make further improvements: Pembroke Hall to be enlarged', 26 August 1938; *Irish Times*, 'Further improvements at Ballsbridge', 4 November 1938.

35. Lyal Collen Papers, *Collen: An Irish Building Family 1782–1992*, p.2.

36. Collen Papers Portadown, Speech by Joe Collen to the Rotary Society, *My Job*, 1971, p.2.

37. Interview with Joe Collen, 14 May 2009.

38. *Ibid.*; Interview with Niall Collen, 5 February 2010.

sewerage scheme for Portadown in 1929, benefiting from a long-term business association with the local Urban District Council.[39] The company also
continued its traditional involvement in church building, securing an important local contract from the Church of Ireland for the extension of St Mark's
church, Portadown. The rector, Rev. G.W. Millington and the select vestry
representing the congregation, wished to extend the nave and build a new
tower for the church that would also serve as a war memorial, replacing the
original tower built by the firm during the 1880s. The vestry awarded the
project to Collen in 1928. This was not due primarily to the firm's extensive local connections, but was apparently a commercial decision as Collen's
tender was the lowest available.[40] The foundation stone for the new tower
was laid on 1 November 1928 by C.F. D'Arcy, wife of Dr Charles D'Arcy, the
Church of Ireland Archbishop of Armagh, at a ceremony attended by David
Collen.[41] The firm built the new tower and the extension to St Mark's on
schedule by the autumn of 1930, so that the archbishop was able to celebrate
the consecration of the expanded nave and tower on the same date exactly
two years later.[42] It is apparent that the firm was not simply relying on local
connections, but on its ability to compete effectively for jobs and to complete
projects in a timely fashion.

Jack Collen became seriously ill in 1937, contracting a chronic skin disease.
While his condition was not life threatening, it restricted his involvement in
the day-to-day running of the business. He spent much of his time in Dublin
during the early 1940s, staying with his uncle David in Kilbarrack House
while recuperating from his illness.[43] Although Jack remained a director, the
management of Collen Brothers in Portadown largely passed into the hands
of his sons Tony and Joe Collen, who were the first members of the fifth
generation to take a key role in running the firm.[44] It was the two younger
men who directed the northern branch of the company through the trials and
opportunities of the Second World War.

39. Collen Papers Portadown, Joe Collen Note, *List of contracts*, 1971, p.1.

40. *Church Booklet, St Mark's Portadown*, February 1992, pp.21–4.

41. *Ibid.* p.68.

42. *Ibid.* p.24.

43. Collen Papers Portadown, *Case for Counsel to advise J.B. Collen*, 1943, p.3.

44. Lyal Collen Papers, Lyal Collen Note, p.1.

'... THE BOYS THOUGHT IT WAS GREAT TO GET AWAY TO LONDON FOR A GOOD TIME ...'

The outbreak of war between the Allies and Hitler's Nazi dictatorship in 1939 had far-reaching implications for Collen in Northern Ireland. The province acquired a new strategic significance during the global conflict, especially when the ports of southern Ireland were no longer available due to the neutrality of the Irish state throughout the war. It was vital for the British war effort to keep sea lanes open during the struggle against German submarine warfare in the Battle of the Atlantic and in the absence of southern Irish bases it became essential to divert the Allied convoys around the north coast of Ireland.[45] The naval and air bases in Northern Ireland therefore assumed much greater importance for the successful defence of Britain. Moreover, approximately 300,000 US troops arrived in Northern Ireland during the war, as the Allied forces were concentrated to prepare for the liberation of Europe.[46] The scale of the military presence gave a dramatic boost to the North's economy and created highly favourable prospects for construction activity in the province. Collen Brothers revived its previous association with the British army on a much larger scale than before. The company secured a substantial volume of business from the army, the Royal Air Force (RAF) and civilian authorities overseeing Air Raid Protection measures in the major cities. Collen played a leading role in building military camps for Allied troops and in the construction of airfields. The firm undertook extensive building work at Aldergrove aerodrome for the RAF.[47] The joinery department was heavily involved in wartime construction, manufacturing timber-framed huts and latrines for Allied troops at the military bases and for POW camps at Gilford, where German prisoners were held. The demands of the wartime projects were so great that the joinery workshop did not provide enough space and its operations extended into the street – the joiners were manufacturing huts and other timber-framed structures in Hanover Street, outside the builders' yard.[48] The company benefited extensively from the surge in construction created by the Allied military presence, but through its role in fulfilling the local capital demands of the war effort.

---

45. Ferriter, *Transformation of Ireland*, pp.448–9.

46. *Ibid.* p.449.

47. Lyal Collen Papers, *Collen: An Irish Building Family 1782–1992*, p.2; Collen Papers Portadown, Joe Collen, *My Job*, 1971, p.2.

48. Interview with Niall Collen, 5 February 2010.

The firm also made a useful contribution to civil defence, particularly in the area of air-raid protection.

Northern Ireland was directly affected by the war: Belfast suffered devastating German bombing raids in April and May 1941, which led to the death of over 1100 people. While civil defence was poorly developed in the province before the raids, the shock of the attack stimulated the Stormont government to give a higher priority to Air Raid Protection (ARP) measures.[49] Collen participated fully in ARP measures, building air raid shelters in the cities and towns throughout the province. The company was also involved in repairing bomb damage to British cities scarred by German attacks during the Battle of Britain. Collen was one of the firms employed by the British government to undertake such repair work and teams of its workers travelled from Portadown to war torn English cities, particularly London, during the early 1940s. Joe Collen's son Niall later heard stories from older employees about their adventures in London at the height of the Blitz: 'I remember Wilfie Doak, the foreman and joiner, telling me years ago that he had gone over there and he was only young at the time and the boys thought it was great to get away to London for a good time even though it was during the war.'[50]

Collen did not usually operate outside Ireland at all and the presence of its workers in English cities highlighted the exceptional demands thrown up by the world conflict. The unusual scale of the wartime projects created problems as well as opportunities. The company's solicitors noted in a confidential legal brief, composed in 1943, that Collen Brothers owed £40,000 in bank debts in 1941–2. This was not a sign of impending financial meltdown but rather an unintended consequence of commercial success; it was attributed particularly 'to the very large amount of contracts on hand for the Government and was necessary because the Government Departments were slow in paying.'[51] It was not entirely clear whether the solicitors were referring to the tardiness of the British government or the administration in Stormont, but they also noted that the debt was greatly reduced by 1943 and that it was considerably outweighed by money owed to the firm, mainly by public authorities. It was a far cry from the bleak economic conditions of the previous decade. The war opened up exceptional commercial opportunities for Collen in Northern Ireland, but also imposed unprecedented pressures on its managers and employees. The

---

49. Ferriter, *Transformation of Ireland*, p.446.

50. Interview with Niall Collen, 14 May 2009.

51. Collen Papers Portadown, *Case for Counsel to advise J.B. Collen*, 1943, p.4.

Demolition of air raid shelter, Northern Ireland, 1946. Courtesy of Niall Collen.

firm's annual general meetings were frequently delayed in the early 1940s due to pressure of work; the company's solicitors noted that the AGM had been held at 'intervals of more than a year. This was due to the abnormal amount of work on hands due to Government contracts.'[52] Collen Brothers operated under extraordinary conditions in Northern Ireland during the war; the firm's greatest concern was not finding work, which came thick and fast, but adapting to an unprecedented volume of wartime public contracts.

The Second World War imposed different demands and limitations on Collen Brothers south of the border. The wartime policy of neutrality and the rigorous measures taken by de Valera's government to avoid military involvement preserved Ireland from the worst ravages of the global conflict. But neutrality could not protect Ireland from the economic consequences of the war or prevent several destructive bombings in Dublin, apparently due to errors by the Luftwaffe. Much of Collen's pre-war business declined during the conflict, as the rationing of essential materials impinged on economic activity and public spending on major domestic projects was reduced. Moreover, the volume of military contracts was inevitably much less significant in neutral Ireland and the firm's activity south of the border proceeded on a more modest scale than in Portadown.[53] Yet the company's activity in Dublin

52. *Ibid.*

53. Lyal Collen Papers, *The Collen Building Tradition*, p.2.

was also shaped by wartime demands, notably the building of air-raid shelters and the reconstruction of housing damaged by occasional German bombing. The threat of aerial bombing on Dublin was particularly acute between 1940 and 1942, when the state moved to mobilize local ARP volunteers and services in different areas of the city.[54] The building of air-raid shelters was a central element of ARP work, which required specialized expertise and personnel that could be provided only by the larger building companies in Dublin. Collen Brothers played a significant part in building air-raid shelters ordered by Dublin Corporation, mainly in the older quarters of the city between the canals.[55] The shelters were usually reinforced concrete structures, built on the surface rather than underground, which were intended to hold up to 110 people each; the shelters were designed to slide on a concrete slab that would allow the structure to move horizontally with the force of a bomb.[56] The company built about ninety air-raid shelters in Dublin between the summer of 1940 and May 1944, over a quarter of the total provided by the Corporation during the war.[57] The shelters alone, however, did not provide complete protection against random incidents of bombing, especially in the absence of adequate air defences, which Dublin lacked throughout the war.

The city's vulnerability to attack from the air was demonstrated with brutal clarity on the night of 30–31 May 1941, when four German bombs fell on Dublin. It was by no means the first German attack that had missed its target, but it was easily the most destructive. Two bombs fell on the North Circular Road and in the Phoenix Park close to Dublin Zoo, but a single bomb wreaked the greatest devastation on the North Strand, where forty-one people were killed, a much larger number were injured and over 500 were rendered homeless.[58] The structural damage caused by the bombing in a densely populated and overcrowded area was equally extensive – over 1700 houses were severely damaged, while of this total sixty-nine were destroyed.[59] The widespread destruction caused by the bombing created an urgent demand for swift short-term repair work. Collen took the lead in undertaking the

---

54. *Irish Times*. 'Air raid precautions', 4 January 1941.

55. Lyal Collen Papers, *Note on ARP*, pp.1–2.

56. *Ibid.*

57. *Ibid.*; River House Archive, Sales Book for Collen Brothers 1900–49, pp.57–9.

58. Lyal Collen Papers, *Note on ARP*, p.3; *Irish Times*, 'German bombs dropped on Dublin: Many killed, injured and homeless', 7 June 1941.

59. *Irish Times*, 'North Strand bombed areas: Damage to 1,700 houses', 1 November 1941.

essential repair work in the North Strand, which consisted mainly of minor repairs to lightly damaged houses and the rebuilding of more severely damaged dwellings.[60] The official in charge of the bomb damage repairs, R.S. Laurie, told the *Irish Times* in November 1941 that 'first-aid repairs' were largely completed, while the 'reinstatement' of damaged houses was still ongoing, with 620 workers from different contractors employed at the bombed site.[61] The company's involvement in the rebuilding operations was concluded by November 1942, when the final payment for its work in the North Strand was received.[62] Although wider plans by the Corporation for a general redevelopment of the area were not implemented in the short term, Collen Brothers played a significant role in the reconstruction of the North Strand following the bombings.

The company's activity south of the border was heavily influenced not only by Ireland's status as a neutral state but also by the economic policies of de Valera's government during the conflict. The Fianna Fáil government sought to encourage tillage at the expense of dairy farming and this approach, combined with the wider policy of economic protectionism, had important implications for the development of flour milling. The government's agricultural policies incorporated an attempt to be self-sufficient in flour, but this did not work out as anticipated. While the new policy direction created a greater demand for flour milling, the move towards self-sufficiency ironically allowed the British company Ranks to secure a dominant role in the Irish flour milling industry.[63] Nevertheless government policy encouraged an expansion in the construction of grain drying and storage plants in the 1940s, which opened up a new stream of business for Collen. The firm undertook its first project for the flour milling industry during the war, building a new grain silo for the Dock Milling company in Barrow Street between 1941 and 1942.[64] The project involved the construction of a reinforced concrete silo, containing twelve storage bins with a capacity to hold about seventy tons of grain each.[65] The job presented some technical difficulty, as a very smooth surface finish was required to the concrete in the silo. Harky's younger son, Lyal, who

---

60. River House Archive, Sales Book for Collen Brothers 1900–49, p.56.

61. *Irish Times*, 'North Strand bombed areas: Damage to 1,700 houses', 1 November 1941.

62. River House Archive, Sales Book for Collen Brothers 1900–49, p.56.

63. Ferriter, *Transformation of Ireland*, p.372.

64. Lyal Collen Papers, Lyal Collen Note, *Grain Silos*, p.1.

65. *Ibid.*

had recently graduated from Trinity College Dublin with a degree in civil engineering, took the opportunity to experiment with his own system for achieving a high-quality finish. This essentially comprised an electric motor deliberately modified to be out of balance and fitted to the shutters on the silo, to vibrate and cause thorough compaction of the concrete immediately behind the formwork, providing a dense material free of imperfections. This method of vibrating the concrete, which later became a well-known technique, was highly innovative in the early 1940s.[66] This method appeared to be successful, but it was not always reliable and was later replaced by newer technology.[67] The project was perhaps less important for Lyal's attempts at technological innovation than for his emergence as a key figure within the firm. The contract was significant too because it pointed the way towards new commercial opportunities: the construction of grain silos became a significant element of the company's business in the post war era.

Grain Silo for Dock Milling Company, Barrow St., Dublin, 1941.
Collen Brothers photo album.

66. Correspondence with Dr. Brian Bond, 30 January 2010.
67. Lyal Collen Papers, Lyal Collen Note, *Grain Silos*, p.2.

The war served as an important formative experience for leading members of the family business during the post-war era. Lyal began a highly successful career in business during the conflict, acquiring management and technical expertise within the firm. Lyal's central role in overseeing the construction of the grain silo at Barrow Street in 1941–2 marked the beginning of a fifty-year career within various incarnations of the company. He was also deeply engaged with the firm's contribution to civil defence, particularly ARP measures in Dublin. Meanwhile, Standish volunteered for the British army, serving as an officer in the Royal Engineers. He gained considerable expertize in the use of bailey or pontoon bridges, temporary structures designed to transport soldiers and munitions across rivers where the original bridges had been destroyed.[68] Standish was not directly involved in the famous D-Day landings by Allied forces on the Normandy beaches on 6 June 1944, but he was certainly familiar with one of the major engineering innovations of the war, the construction of two Mulberry Harbours used during the military operation.[69] The Mulberry Harbours comprised a large number of concrete caissons, which were built on the south coast of England and towed across the English Channel to be scuttled close to the Normandy coast. These provided breakwaters and supported miles of steel roadway, which were used to bring in reinforcements for the initial invasion force and to unload vast quantities of equipment and munitions. Following the success of the D-Day landings, Standish served with a unit of Royal Engineers during the advance by British forces on Germany in 1945 and his final military operation involved throwing a bailey bridge across the Rhine in the last months of the war.[70] Standish's experience as a military engineer would prove invaluable when he returned to Ireland after 1945 to take a leading role in the company. Yet the conflict was even more important as a turning point in his personal life. Standish met his future wife, Claire Wilson, during the war; she was a member of the auxiliary division of the New Zealand Air Force, which had units stationed in Britain. They returned to Ireland together and were married in August 1947 at the Abbey church in Dublin.[71]

---

68. Interview with Martin Glynn, 16 June 2009; Interview with Paddy Wall, 28 April 2009.

69. Ronnie Hoffman, 'It took 200 years to build', *Business and Finance*, vol.17, no.33, 30 April 1981.

70. *Ibid.*

71. *Irish Times*, 'Wedding: Mr. S. Collen and Miss C. Wilson', 19 August 1947.

The Woodside Estate, Portadown, 1948. Courtesy of Niall Collen.

Collen Brothers enjoyed considerable success on both sides of the border immediately after the end of the Second World War. The firm benefited from the efforts of the Northern Ireland administration, and perhaps still more the UK government, to promote reconstruction and development in the post-war era. Collen secured a major contract in August 1946 to build 134 houses on the Woodside estate in Portadown: the Urban District Council accepted the firm's tender on 23 August by six votes to one.[72] The project was the first large-scale public housing scheme completed by the firm in Northern Ireland. Joe Collen regarded it as a vital contract in allowing the company to adapt successfully to post-war economic conditions following the end of the wartime boom in construction.[73] The leading figures of the firm retained vivid memories of the bleak economic environment in the 1930s and the Woodside project gave some assurance that the end of the war would not mark a reversion to the slump of the previous decade. The project foreshadowed the development of major public housing schemes in Northern Ireland during the post-war era, which would have profound implications for Collen Brothers.

72. *Irish Times*, 'Housing scheme for Portadown', 23 August 1946.
73. Interview with Niall Collen, 5 February 2010.

Opening ceremony for new Iveagh Trust flats, Kevin St., 1951.
Collen Brothers photo album.

Yet the provision of residential housing remained only a peripheral element of Collen's activity south of the border. The firm built an extension of the Iveagh Trust housing scheme in the south inner city between 1943 and 1950, providing two new blocks of apartments in Kevin Street.[74] But the Iveagh Trust was the only significant housing development undertaken by Collen in Dublin around this time. The Dublin branch of the company instead became more heavily involved in civil engineering work immediately after the war. Collen completed the site work and recreational facilities for Butlin's holiday camp near Gormanston, Co. Meath in 1947.[75] The first Butlin's camp to be established outside the UK, the holiday village at Mosney opened its doors in 1948. The firm also constructed new bridges at Charlemont Street and Killester for Dublin Corporation during the 1940s.[76] Collen did not, however, neglect major building projects for prestigious

74. River House Archive, *Collen Brothers, List of Accounts 1940–50*, Iveagh Trust account 1943–50.

75. Lyal Collen Papers, *The Collen Building Tradition*, p.2; River House Archive, Collen Photo Album, 1941–51.

76. River House Archive, Photo Album, 1941–51.

clients, completing Sandymount Hall in Ballsbridge for the RDS in 1949.[77] The southern branch of the firm, which had lagged behind its northern counterpart during the war, enjoyed a somewhat higher volume of business in the late 1940s. Yet there was little to choose between the financial performance of the two sides of the company. While the Portadown contracts account showed a turnover of £170,747 in 1949, the turnover for the Dublin account was only marginally higher at £186,817.[78] It was not primarily financial concerns but the culmination of a long-running internal debate among the partners that ultimately led to the division of the firm between north and south at the end of the 1940s.

Top: Butlin's holiday camp, Mosney, 1947. Collen Brothers photo album.
Bottom: Sandymount Hall, Ballsbridge, RDS, 1949. Collen Brothers photo album.

77. River House Archive, Sales Book for Collen Brothers 1900–49, p.82.

78. Collen Papers Portadown, Dublin Contracts Account for year ended 31 December 1949; Portadown Contracts Account for year ended 31 December 1949.

'SO FAR AS I CAN SEE THE WHOLE AFFAIRS OF THE COMPANY WILL HAVE TO BE GONE INTO AND WOUND UP ...' (DAVID COLLEN, NOVEMBER 1922)

The formal separation of the firm into its northern and southern components was seriously considered within the family as early as the 1920s. Following the death of John Collen in 1921, which closely coincided with the partition of the island, the question of dividing the company reared its head for the first time. While the reality of partition provided the context for the family's deliberations, their concerns were not driven primarily by politics, but by financial issues and disagreements concerning the future ownership and direction of the firm; in particular, the question of succession to the founding generation loomed large in an internal debate among the remaining partners. The forty shares held by John Collen at the time of his death (out of a total of 240) were divided in accordance with his intentions, with twenty being allocated to his brother Richard and ten each to his nephews Harky and Jack Collen.[79] The re-allocation of John's stake in the company left Joseph Collen and his two sons in possession of a majority shareholding within Collen Brothers, reflecting their pivotal position in the management of the firm by the early 1920s. But David Collen, the youngest of the original partners, was deeply dissatisfied with such a reordering of the shareholding; he threatened to end his involvement with the firm in May 1922 unless he received a share of his late brother's holdings.[80] A dispute between the partners was avoided when Richard, acting as a peacemaker, agreed to arrange for the transfer of twenty shares, mainly from his own stake, to David; some of these shares were allocated to his son Richard Junior (Dick), who became a director of the firm for the first time.[81] This settlement prevented the break-up of the original partnership, but did not affect the balance of power within the firm established following John's death. David was by no means entirely mollified and his dissatisfaction may have influenced his decision to advocate a formal split within the company later in 1922. He argued that dividing the firm was inevitable, in a letter to his brother Richard on 1 November 1922: 'So far as I can see the whole affairs of the company will have to be gone into and wound

---

79. Collen Papers Portadown, Register of Shareholders, Collen Brothers Limited, 31 January 1922, p.11.

80. Collen Papers Portadown, David Collen to Richard Collen, 12 May 1922.

81. David Collen to Richard Collen, 22 June 1922.

up and then two new companies formed … but as things are at present the power of voting is all on one side, so that anything I might propose would be useless.'[82]

Whatever the merit of David's complaints about the value of his shareholding, he was by no means alone in considering a formal division of the company. His elder brother Joseph investigated the possibility of establishing a new company in Dublin, discussing the idea with Richard in October 1921; Joseph also commissioned a confidential report by Hayes and Sons solicitors in 1922 to examine the prospects for dividing the business.[83] But he soon rejected the idea, not least because the report highlighted the substantial costs involved in such a move. A new company would require considerable initial expenditure on property, stamp duty and solicitors' fees, while facing an increased liability for income tax.[84] Joseph and his son Harky wrote to Richard in December 1922, definitively ruling out a formal split in the company on financial grounds. The two men recommended instead the maintenance of separate accounts and records for the firm's operations north and south of the border, which would still be retained in the Portadown office. They commented that 'you see the difficulties that would arise, as to having to pay income tax on this … We think from what we hear about other Companies in a similar position that it is better not to make any change, except to keep the booking as separate as possible in Portadown.'[85] Their intervention proved decisive, as it underlined that there was considerable opposition among the partners to the radical expedient of a split. The directors kept the registered office of the company in Portadown, while maintaining separate records for the two branches of the business. The internal debate also settled the key issue of the succession in the short-term, confirming the dominant position of Harky and Jack Collen within the firm during the inter-war period. The prospect of dividing the company receded for another two decades.

Yet the internal deliberations of the partners in the early 1920s did not fully resolve the thorny issues of internal succession. Moreover, the debate over the division of the company was postponed rather than resolved. The possibility of a split re-emerged as a point of contention among the directors

---

82. Collen Papers Portadown, David Collen to Richard Collen, 1 November 1922.

83. Joseph Collen to Richard Collen, 14 October 1921; Joseph Collen to Richard Collen, 21 December 1922.

84. Joseph Collen to Richard Collen, 21 December 1922.

85. *Ibid.*

when Joseph Collen died on 25 July 1941.[86] On this occasion the outgoing chairman bequeathed all of his forty shares to his own children and grandchildren, with half being allocated to Harky and Jack Collen, while the remainder was divided between Joseph's daughters and Harky's only daughter Dorothy.[87] Harky and Jack already held a majority of the company's shares even before their father's death and appeared set to reinforce their position on the basis of Joseph's bequest.[88] They bought the shares originally earmarked for their sisters in February 1942, taking possession of three-quarters of Joseph's original shareholding.[89] This arrangement, however, provoked an indignant response from the other surviving directors, David Collen and his son Dick. David was now almost eighty years of age and had retired from active involvement in the family business, but he had lost none of his forcefulness or willingness to challenge other family members; he was determined to promote his interests and perhaps especially those of his son Dick. Jack and Harky Collen respected David's previous contribution to the firm, and Jack in particular enjoyed a friendly relationship with his youngest uncle and was staying with David at Kilbarrack House in 1941–42. Yet the two brothers understandably believed that they were entitled to take over their father's shareholding and were strongly averse to making any concessions to Dick, who had never been involved in the company's work.[90]

The two active directors initially took a conciliatory approach, explicitly recognizing their uncle's position as the senior partner. Harky and Jack ensured that David was appointed as chairman of Collen Brothers at the firm's annual general meeting (AGM) on 8 October 1941. The older man in turn paid an emotional tribute to his late brother Joseph, emphasizing 'His death, I feel very much, and feel lonely being the surviving member of the old Firm.'[91] The initial meeting following Joseph's death appeared to augur well for future collaboration, but David soon came into conflict with his nephews. The differences between the two sides of the family over share ownership re-emerged at the firm's AGM

---

86. *Irish Times*, 'Irish Wills', 13 November 1941.

87. Collen Papers Portadown, *Case for Counsel to advise J.B. Collen*, 1943, p.2. The precise distribution of the shares was: Harky Collen – 10; Jack Collen – 10; Dorothy Collen – 10; 10 between Hannah Sophie Collen, Margaret Beckett and Mary Elizabeth Heron.

88. Register of Shareholders, Collen Brothers Limited, 1936, p.17.

89. Collen Papers Portadown, *Case for Counsel to advise J.B. Collen*, 1943, p.2.

90. *Ibid.* p.3.

91. Collen Papers Portadown, Minutes of Annual General Meeting of Collen Bros Ltd, 8 October 1941.

in Dublin on 10 August 1943. When Harky and Jack sought to have the transfer
of their father's shares into their own names formally included on the shareholders' register, David objected on the basis that under the company's rules
the redistribution of the shares could be decided only by a meeting of the directors.[92] He refused to allow the transfer to take effect and the meeting broke up
without reaching any agreement. Shortly afterwards, David wrote to the secretary, W.T. Church, on 26 August, criticizing the management of the firm and
proposing a division of the company between the two sides of the family:

> As for the last few years the business of Collen Brothers Ltd has not been carried on in
> a satisfactory manner, I think it is time to bring the present state of affairs to an end,
> and in order to do so, I propose that … the total [be] divided into two equal parts.
> One part to go to J.B. Collen and his brother J.H. Collen and the other part to come
> to Richard Collen and myself.[93]

The senior director envisaged that he and Dick would receive half of the
firm's assets, including the East Wall yard, while the remainder, including
the premises in Portadown, would be left to his nephews. It was not entirely
clear whether he hoped for the establishment of two separate companies or a
straightforward division of the firm's assets between the two branches of the
family. It is apparent, however, that David was seeking to reopen the question
of the succession to the founding generation within Collen Brothers, which
had apparently been resolved almost twenty years earlier. He was not recon-
ciled to the pivotal position secured by his nephews within the company and
hoped to secure a share of the business for his son, if necessary by dividing
the firm. Once again disagreement over the distribution of a key proprietor's
shareholding had provoked a debate about the future of the firm itself.

David Collen's proposal raised fundamental issues for Collen Brothers,
creating the real prospect that the company would cease to have a presence in
Dublin and even that it might not survive in its existing form. He certainly
hoped that Collen Brothers would survive, but was vague about how this would
be achieved or whether the firm would maintain a presence in Dublin: 'If J.B.
Collen and J.H Collen agree to the arrangement I have proposed, it can be
brought [about] without much trouble, and then they and their sons can carry
on to suit themselves.'[94] But Harky and Jack had no hesitation in rejecting

---

92. *Case for Counsel to advise J.B. Collen*, 1943, pp.3–4.

93. Collen Papers Portadown, David Collen to the Secretary, Collen Bros Ltd, 26 August
1943.

94. *Ibid.*

their uncle's initiative. W.T Church conveyed their response to David on 9 September, noting that 'they would not agree under any consideration that the assets of the Company would be divided into two equal parts.'[95] The two men were seriously alarmed by David's proposal, not least because they had been managing the business on both sides of the border for the previous two decades. Jack Collen sought legal advice from solicitors in Portadown, T.D. Gibson and Co.; his legal advisers drew up a detailed brief, which vividly illustrated the concerns of the two active directors. Harky and Jack regarded David's proposal to divide the assets of the business as 'grossly unfair' in the light of their central role in managing the company.[96] They were also concerned about the position of their sons, who were already employed in the business but might be unable to become shareholders unless the dispute was resolved.[97] The relations between the two groups of directors in the early to mid-1940s were marked by a degree of mutual acrimony that was unusual in the history of Collen Brothers. The directors were divided into opposing camps, with each holding an effective veto over the objectives of the other; the two brothers prevented any move by their relatives to divide the company, but equally David and his son acted to block the formal reallocation of Joseph Collen's shares. The company's annual general meetings were characterized by gridlock over a two-year period, with no decisions of any consequence being made between 1942 and 1944.

Yet both parties drew back from further confrontation, seeking other means of resolving their dispute. Harky and Jack sought in June 1944 to buy out their relatives' stake in the company: they offered £15,000 for the shares held by their uncle and cousin, with the stipulation that the offer was subject to the deduction of any debts owed by their relatives to the company.[98] This was an important sting in the tail, as Dick Collen owed over £2800 to the firm in 1943. While David and his son were willing to consider the possibility of selling their shareholding, they were dissatisfied with the financial terms offered by the other two directors. But while the issues involved in the dispute were not resolved in a clear-cut fashion, neither party wished to prolong the conflict indefinitely. When W.T. Church asked Harky on 6 September 1944 to

---

95. Collen Papers Portadown, W.T. Church to David Collen, 9 September 1943.

96. *Case for Counsel to advise J.B. Collen*, 1943, p.5.

97. *Ibid.*

98. Collen Papers Portadown, T.D. Gibson & Co. to Wheeler & McCutcheon, 19 June 1944, Wheeler & McCutcheon to T.D Gibson & Co., 10 July 1944, T.D Gibson & Co. to Wheeler & McCutcheon, 11 July 1944.

clarify whether an AGM should be convened in the short-term, Harky immediately urged him not to do so, having heard that his uncle was ill: 'I don't
think it would be advisable to send out notices … Mr D.C. is not well and
would not be able to attend.'[99] The younger man took a conciliatory line,
seeking to avoid deepening the rift with his relatives. The dispute among the
directors gradually petered out, due in part to the reluctance of most family
members to pursue the conflict aggressively. Several key protagonists were
also removed by the passage of time. David Collen died on 4 February 1945.
An obituary in the *Irish Times* noted that he was the final survivor of the
original founding partners of Collen Brothers and also paid tribute to his
prominent role within the Standing Committee of the Methodist church.[100]
Dick remained a shareholder of Collen Brothers for the rest of his life but
was not directly involved in the management of the company, and he died
on 6 April 1948.[101] Harky and Jack Collen became the sole proprietors of the
company virtually by default during the late 1940s.

The resolution of the acrimonious but relatively brief internal dispute facilitated the emergence of a consensus on the future direction of the company.
The question of dividing the firm between its northern and southern components came to the fore once again shortly after the end of the Second World
War. The wider political backdrop certainly presented a logical rationale for
the formal division of the firm. It was evident by the late 1940s that partition was an enduring reality, not a temporary phenomenon. The decision by
the inter-party government headed by John A. Costello to proclaim Ireland
a republic, which was formally inaugurated in April 1949, reinforced the
existing division between north and south, regardless of the anti-partition
rhetoric of some leading ministers. The declaration of the republic was closely
followed by the Ireland Act, passed by the British Parliament in 1949, which
provided a legislative guarantee that Northern Ireland would not be excluded
from the United Kingdom against the will of the devolved parliament.[102] Yet
the political context was not the primary influence on the family deliberations in Dublin and Portadown, which led to the emergence of two separate
companies on each side of the border. It was internal pressures for change,
rather than external circumstances, which led to the division of the firm.

---

99. Church to J.H Collen, 6 September 1944; J.H. Collen to Church, 7 September 1944.

100. *Irish Times*, 'Obituary: Mr David Collen', 6 February 1945.

101. Collen Papers Portadown, S.B.I Abbott & Co. to W.T. Church, 19 December 1949.

102. FSL Lyons, *Ireland Since the Famine* (London, 1972), pp.567–8.

An ongoing generational transition within the southern branch of the company was the most significant influence on the decision. The initiative for the division of the firm came from the younger members of the family in Dublin. Standish had returned from the British army to join the firm during the late 1940s, while Lyal was already prominently involved in the business by 1945. Joe Collen believed that the move occurred at the instigation of his cousins in East Wall: 'They just came up and the thing happened, most unexpectedly to me. I remember Standish being the main one.'[103] There is no doubt that Lyal and Standish favoured a division of the firm, with Standish acting as the leading advocate of change.

It is easier to identify the authors of the move than to determine their motivations with complete certainty. The move to divide the firm was not dictated by financial pressures or poor commercial performance. Significantly, both branches of the firm were in credit and there was no great difference between Dublin and Portadown with regard to profits and commercial performance. The unified company's balance sheet up to 31 December 1949 indicated that the Dublin account showed a net profit of just over £11,040, while the Portadown account displayed a slightly smaller profit of £8272.[104] While the Dublin branch showed a marginally stronger performance, it was far from sufficient to provide a case to divide the firm and in fact the overall balance sheet underlined that both sections of the company were performing strongly. Yet it was apparent that the structure of decision-making at board level no longer reflected the realities of the business on the ground, which had increasingly evolved over time into two separate units. The firm's procedures, which maintained the registered office of the company in Hanover Street and required that annual general meetings should only be held in its offices in Portadown, were increasingly cumbersome and awkward for the managers in Dublin.[105] Moreover, the events of the early 1940s graphically demonstrated that formal decision-making at board level had the potential to grind to a halt if a serious disagreement emerged among the directors, even when this had little to do with the actual management of the firm. It is likely that Lyal and Standish were influenced less by narrowly financial considerations and more by a desire to establish an independent company under their direction.

---

103. Interview with Joe Collen, 14 May 2009.

104. Collen Papers Portadown, Collen Brothers, *General Profit and Loss Account for the Year ended 31st December 1949.*

105. *Case for Counsel to advise J.B. Collen,* 1943, pp.3–4.

While their initiative certainly caused some surprise among the family members in Portadown, especially Joe Collen, there was no indication of any opposition to the division of the company. Standish and Lyal undoubtedly took the lead in making the case for a formal separation between the two branches of the business, but it is apparent that their pressure did not meet any significant resistance on the part of the senior directors. The decision ultimately rested with Harky and Jack Collen, who had successfully resisted efforts to divide the company only a few years earlier. Yet on this occasion both senior figures endorsed the initiative. The family reached a consensus on how to divide the firm towards the end of 1949, which involved the establishment of a new company in Dublin and the transfer of Harky Collen's shares in the existing firm to his relatives in Portadown. Harky was deeply involved in negotiating the details with his brother, writing to Jack on 13 December 1949 to agree the terms for dividing the business. Harky told his brother that 'we have got the new company, Collen Bros (Dublin) Ltd ready for registration' and suggested that a meeting of the directors should be held in Portadown on 22 December to make the final arrangements for the division of the original firm.[106] He also gave his brother practical advice on how to manage the transformation of the business. Harky intended to allocate the large majority of the shares in the new company to his sons and advised Jack to follow his example as 'this eliminates larger death duties at a later date.'[107] W.T. Church replied to Harky two days later, conveying Jack Collen's favourable response: 'Mr. J.B. Collen had a conversation with Mr Bright, Solicitor, and consider [sic] everything all right, and that there should be a conveyance from the old Company to you, of the Dublin Property...'[108] The close collaboration between the two brothers was crucial in smoothing the passage for an amicable separation between the two halves of the business.

The long-delayed division of the firm became a reality in December 1949. The final directors' meeting of the all-Ireland company was held on 22 December at the offices of Collen Brothers in Hanover Street, Portadown, to ratify the terms of separation. As the principle of dividing the business had already been accepted, the key issues to be finalized by the directors were the distribution of the firm's assets and the transfer of shares. The assets of the firm consisted

---

106. Harky Collen to Jack Collen, 13 December 1949.

107. *Ibid.*

108. W.T. Church to J.H. Collen, 15 December 1949. Sidney Bright was the family's solicitor in Portadown, who was also Tony Collen's father-in-law.

mainly of property in Dublin and Portadown, as well as a considerable quan-tity of plant and machinery.[109] The company owned a builders' yard, offices, and workshop in Hanover Street; its profile south of the border was very similar, with its most significant asset being the yard and premises at East Wall Road. There was also the less tangible but real advantage of the accumulated goodwill built up by the firm over three generations. Both of the newly inde-pendent companies would seek to capitalize on such goodwill by retaining the established brand name of Collen Brothers.

The directors approved the transfer of all the firm's property and plant in the Irish state to the new company in Dublin from 1 January 1950.[110] The ownership of the existing firm was transferred completely to Jack Collen's family in Portadown, with his two sons Tony and Joe emerging as the domi-nant shareholders. Harky transferred his entire shareholding in the old firm to his nephews, who also inherited the shares of David and Dick Collen in line with their wills.[111] Moreover, Jack Collen, following Harky's advice, also took the opportunity to transfer most of his own shares to his sons, who were appointed as directors of the firm for the first time.[112] Joe and Tony Collen became equal shareholders in Collen Brothers, confirming their key position within the business. The division of the traditional firm also marked the culmination of a transition to a new generation on both sides of the border. The directors' meeting settled the outstanding issues and completed the formalities of separation, paving the way for each side of the business to go its own way.

The southern members of the family lost no time in establishing their new company. Collen Brothers Dublin, Ltd. came into existence officially on 23 December, the day after the fateful meeting in Portadown. The new firm was established as a limited liability company based in the recently proclaimed Republic of Ireland. The most immediate purpose of the company was set out in Article 3 of its original memorandum of association, which stipulated that Collen Brothers Dublin would acquire or take over from 1 January 1950 'the entire premises, business, plant, stock-in-trade and goodwill and every other property both real and personal of Collen Brothers Limited in respect of the business of General Contractors and Builders at present carried on in the

109. Collen Papers Portadown, Collen Brothers, *Balance sheet as of 31 December 1949.*
110. W.T. Church to J.H. Collen, 15 December 1949.
111. Register of Shareholders, Collen Brothers Limited, 1936, 1 January 1950, pp.21–2.
112. *Ibid.*; Church to J.H. Collen, 19 December 1949.

Republic of Ireland.'[113] The directors of the new company were Harky himself and his eldest son H.J. Desmond Collen, as well as Standish and Lyal.[114] There was a clear understanding from the outset that each company would operate in its own jurisdiction and that neither would compete against the other. The separation was accomplished in a decisive and clear-cut fashion, involving a precise delineation of operations and a definite division of plant and property between the northern and southern branches of the business. Yet it was also an amicable separation, with each branch of the family recognizing the right of the other to pursue their own interests independently.

The division of the firm in 1949 was not a bolt from the blue, but a recurring issue which first arose almost a generation before it finally became a reality. The question of dividing the firm became inextricably entangled with the central issues of ownership and succession twice in a generation. While there was little or no contention over who managed the company, significant disagreement emerged among the directors over the ownership of shares and ultimately control of the firm in the 1920s and again in a more dramatic fashion during the early 1940s. No agreement on the future structure of the firm was possible while these key issues remained unresolved: it was only when the issues of ownership and control of the firm were settled beyond doubt that the division of the business became a practical option. While the pressure for change came from younger members of the family, the close relationship between Harky and Jack Collen greatly facilitated the agreed division of the company. The family directors and managers reached a consensus on the future direction of the firm towards the end of the 1940s, implementing a definitive but amicable separation of the business between north and south.

113. River House Archive, *Memorandum of Association of Collen Bros. (Dublin) Limited*, p.1.
114. *Ibid.* p.11.

FOUR

An Era of Expansion

Collen Brothers, Dublin, shared many features with its all-Ireland predecessor, reflecting a distinctive family tradition that dated back to the early stages of the Industrial Revolution. The firm began this period as a traditional building company, relying heavily on repeat business from loyal clients or traditional avenues of activity in dismal economic conditions. But the fortunes of the company were transformed during the post-war era. The scale of its activity expanded dramatically between the early 1960s and the late 1970s as the firm took full advantage of the opportunities offered by the policy of economic development driven by Seán Lemass and key officials such as T.K. Whitaker. Collen also extended the scope of its business, developing a far wider range of interests than before, including industrial building projects, large-scale civil engineering work and property development. The company's internal structure and business practices evolved over time, incorporating a significantly enhanced role for managers and directors who were not members of the Collen family. Yet the extent of the transformation should not be overstated. Collen remained above all a family business and the authority to make key decisions rested firmly with the leading members of the family. Moreover, the company's business model retained a strong vein of caution and indeed conservatism, which was modified but by no means abandoned during an era of real if uneven economic expansion.

All four of the original directors of Collen Brothers Dublin Ltd were family members – Harky Collen and his three sons – and the newly constituted company remained under the full ownership and control of the family. The Articles of Association, effectively the founding charter of the company, stipulated that Collen Brothers was a private company with a limited membership consisting initially of the four family shareholders.[1] The Articles of Association imposed restrictions on the transfer of shares, noting that any members who wished to sell their shares were obliged to transfer them to other members. While the sale of shares to non-members was theoretically permitted in the absence of willing buyers among the existing membership, Article 20 gave the directors a veto over any such transaction: 'the Directors may decline to register any transfer of shares to a person of whom they do not approve'.[2] The first meeting of the directors on 28 December 1949 agreed that each man would initially take an equal stake of one ordinary share in the new company.[3] The directors took up further shares over the following two years, with Lyal and Standish securing the majority shareholding in the firm by agreement with their relatives.[4] Harky, who was over seventy years of age in 1949, became chairman of the board and was recognized in the original Articles of Association as the 'Governing Director' of the company.[5] But Standish and Lyal Collen were the dominant figures in managing the business from the outset. The role of the two men was formally recognized in 1961, when the directors agreed to adopt a formal management structure for the first time. Standish and Lyal Collen were appointed as joint managing directors for a ten-year period from 3 April 1961, with the approval of the other family shareholders: at the same time both men gave interest-free loans to the company, amounting to £4000 each, to increase its working capital.[6] The agreement formalized the transition to a new generation, which had occurred, in practice, over a decade before.

All the key decisions taken by Standish and Lyal in this period reflected the central place of the family in the organization of the business. When

---

1. Collen Brothers (Dublin) Limited, *Articles of Association*, pp.11–15.

2. *Ibid.* pp.19–20.

3. Collen Brothers, Minute book, Minutes, First Meeting of Directors, 28 December 1949, pp.1–3.

4. Minutes, Meeting of Directors, 31 October 1951; Minutes, Meeting of Directors, 18 December 1951.

5. Collen Brothers (Dublin) Limited, *Articles of Association*, p.28.

6. Minutes, Meeting of Directors, 28 March 1961, pp.45–6.

Harky died in March 1963, his widow Mary Arnott Collen was immediately appointed to take his place, where she served on the board for two years alongside all three of her sons.[7] Harky's eldest son Desmond also served as a director of Collen Brothers, attending board meetings regularly, but did not take an active part in running the company. Desmond remained a quiet long-term shareholder until his death in 1975.[8] Standish and Lyal alternated as chair of the board between 1963 and 1966, until they decided to fill the office on a permanent basis. Standish was appointed as chairman of the company in September 1966, retaining the position until his retirement almost thirty years later.[9] The generation from the 1950s to the 1980s would be an era of far-reaching transformation within Collen Brothers and more widely for Irish society, but the considerable changes within the company would occur on terms decided by Standish and Lyal Collen.

Standish and Lyal Collen. Lyal Collen Papers.

7. Minutes, Meeting of Directors, 5 April 1963, p.54.

8. *Irish Times*, Notice, 15 May 1975.

9. Minutes, Meeting of Directors, 21 September 1966, p.77.

The Collen family directors traditionally relied upon a few key collabora-tors, who would undertake important administrative responsibilities but were not directors or shareholders within the firm. This was the role fulfilled during and after the Second World War by Harold Townsend, who joined the Dublin branch of the firm as a senior manager in 1940 and acted as secretary of the new company from December 1949.[10] Paddy Wall, who entered the accounts department of the company in August 1961 and later served as company secre-tary in a very different era, recalled that Townsend's title did not do justice to his wide-ranging responsibilities: 'He would have been Standish Collen's right hand man. Now he would have been secretary, wages clerk, legal, administra-tive [assistant], anything to do with the non-technical side of the business, Harold Townsend did it.'[11] Townsend was effectively a civilian adjutant to Standish and remained the principal (and often the only) administrator of the firm for over a decade. The board of the company formally acknowledged their debt of gratitude to Townsend following his death in November 1959, recording a tribute in the minutes 'expressing the Directors' appreciation of the manifold services he has rendered to the Collen family and Firm since 1940.'[12] It was significant that the directors chose to highlight Townsend's services to the family, underlining their distinctive world view in which the fortunes of the family and the firm were closely intertwined. Harold Townsend undoubtedly made a vital if low-key contribution to the stable running of the company. His eclectic job description and limited formal influence made him an ideal collaborator for the directors and a characteristic figure in a tradi-tional family business.

Yet the rapid expansion of the company from the late 1950s inevitably demanded changes in its traditional business practice, not least due to Collen's increasing success in taking on large-scale industrial building and civil engi-neering projects. The growing numbers of employees required to meet the demands of new contracts meant that a single individual could not expect to manage the administrative side of the firm, as Harold Townsend had done in a previous era.[13] His successor as company secretary, Thomas Boyd, also took on a considerable range of responsibilities, including recruiting and

---

10. Minutes, First Meeting of Directors, 28 December 1949, pp.1–3; Minutes, Meeting of Directors, 26 November 1959.

11. Interview with Paddy Wall, 28 April 2009.

12. Minutes, Meeting of Directors, 26 November 1959.

13. Interview with Paddy Wall, 28 April 2009.

interviewing new staff, but the directors began to delegate to a wider range of senior managers for the first time from the 1960s. Similarly the increasing scale and complexity of the civil engineering business demanded specialized expertise, which meant bringing in highly qualified engineers and architects. An era of economic expansion brought a significant delegation of managerial responsibility to engineers, architects and managers drawn from outside the family. The new generation of managers were usually long-term employees of the firm, who were carefully selected for their expertise in particular areas but also their commitment to the business model and values of the company. The directors moved judiciously to delegate greater authority to a small number of key managers. Standish and Lyal Collen established a new landmark for the company in 1964, when they appointed a director from outside the family for the first time: a board meeting attended only by the two men on 20 April 1964 resolved that 'Arthur Alexander Harris be co-opted a Director of the Company…'[14] Archie Harris, who had served with Standish in the Royal Engineers, joined the company in 1947 and was its most senior manager by the mid-1960s.[15] He was a highly respected figure within the company, whose elevation occurred after a prolonged period of service. Harris was deeply involved in the day-to-day administration of the company and also took charge of several industrial building projects, including the reconstruction of Roches Stores in the 1960s.[16]

The proprietors made another key appointment two years later, which had even more profound consequences for the company. John W. Griffin, who was another civil engineering graduate of Trinity College, enjoyed considerable engineering expertise in harbour works, working first for Dublin Port and Docks Board and later for the British Admiralty.[17] Griffin joined Collen in the mid-1950s and was appointed as a director in October 1966, once again following a considerable period with the firm.[18] He played a crucial part in the company's growing involvement in major civil engineering projects, especially the undertaking of large-scale harbour works in Dublin Port. The promotion of Archie Harris and John Griffin was a watershed for Collen Brothers, marking a significant break with traditional practice. It initiated a process of

---

14. Minutes, Meeting of Directors, 20 April 1964.

15. *Irish Times*, 'Mr A.A. Harris', 18 January 1964.

16. Interview with Paddy Wall, 28 July 2009.

17. Correspondence with Dr Brian Bond, January 2010.

18. Minutes, Meeting of Directors, 17 October 1966.

gradual change within the company and established a precedent that helped to smooth the way for greater managerial delegation to directors drawn from outside the family.

The company also experienced long-term changes in this period which formed part of wider technological developments within the construction industry. The most obvious change from the early part of the century was the replacement of the previously ubiquitous horse and cart with motor transport in carrying materials to sites. This trend was already well underway by the time the firm was divided in two. The accounts for the firm's business in Dublin in 1949, just before the establishment of the two separate companies, catalogued the value of contracts relying on new and old forms of transport: 'Motor Work £4817 18 6; Horse Work £820 6 1'.[19] The unsentimental logic of the balance sheet underlined that the horse was already losing its traditional place. Yet the replacement of horse-drawn transport with lorries was a gradual process, which was not fully completed until the 1960s. The firm still employed carters with four horses to draw materials to sites in 1959, while using six lorries for heavier loads.[20] But Jerry O'Leary, who joined the company as a lorry driver in 1959, recalled that 'when the big jobs came up, you know, the horses wouldn't be able to cope with them'.[21] The transition to motor transport was largely completed early in the following decade, although the firm still used horses as late as 1961, as Paddy Wall remembered: 'We had two carters, with four grey horses operated by the brothers, John and Thomas McCormack.'[22] The increasing demands of major building and civil engineering contracts meant that horse-drawn transport became a thing of the past. The increased mechanization of the business was an unmistakable trend in this period, with the introduction of more powerful equipment and machinery for building work. The directors' meetings from the early 1950s showed an increasing concern to invest in modern plant, including bulldozers, dumpers and cranes, sometimes balanced by a desire to limit expenditure in difficult economic circumstances.[23]

---

19. Collen Brothers, *Dublin Profit and Loss Account for Year ended 31st December 1949* (note figures given in pounds, shillings and pence).

20. Interview with Jerry O'Leary, 22 July 2009.

21. Interview with Jerry O'Leary, 22 July 2009.

22. Interview with Paddy Wall, 28 April 2009.

23. Collen Minute book, Notes of meeting, 9 February 1955; Notes of meeting, 23 March 1955; Notes of meeting, 3 November 1955.

Four grey horses on holiday in Streamstown. Courtesy of Paddy Wall.

The firm began to expand its operations at East Wall even in the relatively unfavourable conditions of the 1950s. Collen established a fitters' shop at the East Wall yard between 1955 and 1957.[24] The company also expanded its joinery shop, which pre-dated the formal division of the firm; the joinery department, like its counterpart in Portadown, manufactured timber panels and moulds for different types of jobs. The joinery shop was a vital part of the company's operations, providing all the timber products for every contract undertaken by Collen Brothers over the next generation.[25] The joinery department produced exceptionally high-quality work, ensuring that the firm never bought joinery while it was in operation.[26] Indeed Collen retained its joinery shop well into the 1990s, long after most contractors had changed to a policy of buying in joinery on a competitive tender basis.

The long-term development of Collen Brothers was greatly influenced by another key innovation in the 1960s, namely the establishment of a distinct design unit within the company. The firm had been involved intermittently in designing buildings as well as constructing them since the early 1900s, but

24. Collen Minute book, Notes of meeting, 3 November 1955; Notes of meeting, 29 July 1957.

25. Interview with Paddy Wall, 14 January 2010.

26. Interview with Jimmy Small, 6 July 2009.

Collen Brothers' truck at East Wall Road; the fitters' shop is in the background.
Collen Brothers (Dublin) photo album.

the design function had traditionally been a back-up element to the main business of building.[27] The design function took on much greater significance from the 1960s, as the company undertook an increasing range and variety of projects. Lyal Collen, who was strongly committed to innovation in civil engineering and architectural design, took the lead in creating an in-house design office. It was a small-scale operation initially, consisting of only two staff, Ronnie McDowell, the firm's senior engineer in the 1960s, and Chris Lyons, an architect who joined the firm in 1965.[28] But the design office expanded in the late 1960s, as the firm attracted several highly qualified civil engineers and structural designers, largely because Collen provided opportunities to engage in design as well as construction. Des Lynch, who joined the company in 1968 from Dublin Corporation, emerged as a leading member of the enlarged design function: he later became head of the design office in the 1970s and managing director of the company itself in 1984.[29] Dr Brian Bond joined the company in 1970, having completed a Ph.D in civil engineering in Trinity College in 1965 and worked for five years with the Goulding group. His move

27. Interview with Martin Glynn, 16 June 2009.

28. Interview with Chris Lyons, 6 July 2009.

29. Interview with Des Lynch, 16 June 2009.

The joinery shop, East Wall Road, during the 1950s

to Collen occurred at least in part because he wanted to be involved in both design and construction, particularly of civil engineering works. His Ph.D was awarded for research carried out in geotechnical engineering and this proved to be of particular value to Collen in several of its civil engineering projects over the following years.[30] He recalled his favourable impression of the company's innovative design and build approach:

> It was, and still is, most unusual, especially in Ireland, for a contractor to have its own in-house design office and it put the company in a unique position to undertake contracts on a design-and-build basis. In the traditional system, a consultant designs the job for the client and then it goes out to tender and the client employs a contractor … to build what the consultant has designed. But there can be problems with this system. Firstly, you get split responsibilities and if problems arise later there can be great arguments about who is to blame. Also consultant designers may design things that are hard to build or they change the design during the construction so that extra costs come in and this can lead to contractual rows developing; so potentially there is a huge advantage in getting one company to do both the design and the construction.[31]

30. Interview with Dr Brian Bond, 30 April 2009.

31. *Ibid.*

The establishment of a strong design function within the firm enhanced its collective technical expertise and gave Collen Brothers considerable scope to undertake design and build contracts, in which the company took overall responsibility for the completion of a project, from obtaining planning permission through designing the structure to the process of construction. Such collective willingness to perceive design and construction as an integrated whole proved an invaluable asset to Collen, facilitating the company's involvement in the development of industrial estates and underpinning its ability to deliver large-scale marine engineering projects on a design and build basis. This development of the design function reached its logical conclusion when the board agreed in July 1969 to investigate the possibility of establishing 'a new Company for design purposes'.[32] Collen Design Ltd was duly incorporated in September 1970, as a limited company under the control of the directors of Collen Brothers – the new entity never traded independently but served as a vehicle for the design office.[33] Collen was exceptional among Irish building companies in developing an in-house design office and in due course setting it up as a separate limited company. The role of the design unit at the time was primarily to undertake design and build contracts for the firm itself rather than independent design work for private clients.[34] The design office later became a more independent operation and, while still providing the design and build capability, also offered design and project management services directly to clients who employed other building contractors. The emergence of a coherent design function as a central element within the firm proved to be a highly successful innovation. While the scale of the design office fluctuated according to changing economic conditions, it remained an enduring feature of the business into the twenty-first century.

The process of change gathered pace in the 1970s, with the appointment of new directors not bearing the Collen name and the emergence of a recognizable senior management team for the first time. Stratton Sharpe, an experienced civil engineer who was deeply involved in the firm's industrial building programme, was appointed to the board in October 1971.[35] Sharpe remained a director for only three years, resigning from the board when he left the employment of the company itself in 1974, taking up a new post with Crampton.[36]

32. Minutes, Meeting of Directors, 29 July 1969.

33. Minutes, Meeting of Directors, 18 September 1970.

34. Interview with Chris Lyons, 6 July 2009.

35. Minutes, 21st AGM of Collen Brothers (Dublin) Ltd, 5 October 1971.

36. Minutes, Meeting of Directors, 31 July 1974; Minutes, Board meeting, 4 December 1974.

But other senior engineers or design specialists were soon elevated to the board. Des Lynch became a director in January 1977, while Brian Bond joined the board in September 1978.[37] Another break with tradition came with the elevation of the first managing director drawn from outside the ranks of the family. John Griffin became managing director of Collen Brothers on 31 July 1974, in a move instigated by Standish and Lyal Collen.[38] The board of directors included as many managers and professional engineers as family members by the late 1970s; it was a striking change for a firm that had not appointed a director from outside the family for almost a century.[39] The opening up of the board to new members testified to an ongoing process of organizational renewal within the firm, as the proprietors integrated valuable managerial and professional expertise into high-level decision-making.

The changing composition of the company's management meetings, which were distinct from the formal meetings of the board, also provided a reliable indication of the growing influence of managers from outside the family. Collen Brothers, Dublin, held regular management meetings from 1950 onwards, which were attended initially only by the Collen family directors (usually with the exception of Desmond Collen) and Harold Townsend.[40] But by 1970 the company's management meetings included not only Standish and Lyal Collen, but also Griffin, Harris and other key figures, who were not themselves members of the board, but performed essential functions within the firm. The regular attendees included Tom Boyd until his retirement in 1971, William Webb who acted as financial controller before succeeding Boyd as secretary, and the long-serving company accountant, William McCullough.[41] This group gradually came to constitute a small core management team, which was deeply involved in considering and implementing most key decisions from the late 1960s through the following decade, although it was not given formal status within the firm and certainly did not act independently of the proprietors. The composition of the senior managerial cohort changed over time, due to retirements and new appointments in the 1970s, but its emergence marked an important evolution in the internal dynamics of the

37. Minutes, 28th AGM of Collen Brothers (Dublin) Ltd., 14 December 1978; Companies Registration Office, Certificate No. 13150/54, Form No.9, *Notification of Change of Directors*, 19 April 1977.

38. Collen Minute book, Minutes, Meeting of Directors, 31 July 1974.

39. Hoffman, 'It took 200 years to build', *Business and Finance*, vol.17, no.33, 30 April 1981.

40. Collen Minute book, Notes of meetings, 25 January 1950–24 January 1966.

41. Collen Minute book, Notes of meetings, 31 January 1966–17 September 1973.

company. It is apparent that senior managers and other key professionals exerted a far more pervasive influence on the fortunes of the company in the 1970s than they had a generation earlier.

The expansion of the board and sustained delegation of managerial functions coincided with the departure of the man who had played an essential, if understated, part in a period of far-reaching change. Archie Harris retired in December 1978, following over a generation as a key figure within the company. His prominent role in managing a wide variety of industrial building projects was widely acknowledged within the firm, while his long service received the greatest attention in the tribute paid to him by the other directors on his retirement.[42] Yet perhaps Harris' most lasting achievement was his invaluable contribution to the unexciting but vital process of organizational renewal; Harris' managerial skill and his status as Standish's trusted collaborator helped to pave the way for an opening up of the firm's structures to professional managerial expertise. The appointment of non-family executives to the board, unprecedented before his promotion in 1964, was a mundane reality of the business a decade later, much to the benefit of the firm.

Yet the authority of managers and other professionals within the firm should not be overstated. While the era following the Second World War undoubtedly saw far-reaching changes within the company, there was also an even stronger level of continuity, reflecting the central place maintained by the family proprietors within the business. Standish and Lyal formed an executive core at the heart of the company, securing expertise from directors drawn from outside the family and delegating a great deal of important work to their managers, but ultimately retaining the final authority to take all key decisions. John Griffin was an influential figure in the development of the company for over two decades, but throughout his tenure as managing director he was obliged to secure the agreement of the family directors for decisions to tender for contracts or undertake significant expenditure. It was Standish, rather than successive managing directors drawn from outside the family, who was the dominant figure in the management of the firm and his approval was indispensable for any major initiative taken by the company. Brian Bond believed that 'Standish was very much the boss. John used to have to go over and chat to Standish to do anything we wanted to do.'[43] Moreover, the interests of the family were deeply intertwined with the development of

42. Collen Minute book, Minutes, Meeting of Directors, 14 December 1978.

43. Interview with Dr Brian Bond, 30 April 2009.

the firm. The company remained firmly under the control of the family, and when the three surviving partners periodically increased the share capital of the company, the new shares were usually retained by the family directors or allocated to other members of the family. The sole exception to this pattern occurred in February 1974, when the partners authorised a new issue of bonus shares. The large majority of these shares were distributed among the close relatives of the directors, including Claire Collen and the children of both Lyal and Standish.[44] But a small proportion of the bonus shares were allocated to Arthur Jolley and Gerard Mahoney, both close friends of Lyal, who held shares in trust on behalf of his immediate family.[45] The predominant position of the family within the business was underlined in September 1978, when Standish's eldest daughter, Diana Collie, was appointed as a director. Diana, who also worked for the company, served on the board until 1984.[46] A highly favourable profile of Collen Brothers, which was published in *Business and Finance* in April 1981, concluded that 'the power is still firmly lodged with the family'.[47] Few employees would have dissented from this assessment, which accurately underlined the key position retained by family members even as the company adapted to far-reaching economic and technological change.

It was fortunate therefore that the two active family directors generally got on well, despite very considerable differences in outlook and temperament. Lyal was the more adventurous of the two men and showed a zeal for innovation. He was favourably disposed towards expanding the company and usually enthusiastic about seeking out new sources of business. Standish was more cautious, pragmatic and focused on the details of management.[48] The two key directors evolved a complementary division of labour over time. Lyal concentrated on the sales and marketing aspects of the business, working to secure clients and bring in contracts – he proved remarkably effective in acquiring work for the company, not least because he enjoyed a wide range of connections in the Irish business world. Standish, meanwhile, was more deeply engaged in managing the business, overseeing the running of the company and the implementation of its contracts. Paddy Wall later commented that 'they were very different people; it was a remarkable

---

44. Collen Minute book, Minutes, Meeting of Directors, 25 February 1974.

45. *Ibid.* pp.1–2.

46. Minutes, Meeting of 28th AGM of Collen Brothers (Dublin) Ltd., 14 December 1978.

47. Hoffman, 'It took 200 years to build', *Business and Finance*, vol.17, no.33, 30 April 1981.

48. Interview with Paddy Wall, 28 April 2009; Interview with Martin Glynn, 16 June 2009.

relationship though because they worked well together.'[49] Standish Collen himself told *Business and Finance* in 1981 that 'During the thirty odd years my brother and I ran this company jointly … I think we only had about three rows.'[50] Sadly for future researchers, Standish was much too shrewd to elaborate on what the particular rows were actually about. It would not be surprising, however, if occasional friction arose between the two brothers as a result of real differences in their business outlook. Yet it is apparent that such friction was kept to a minimum so that the qualities of the two men tended to complement each other. The most eloquent testimony to the strength of their professional relationship was the consistent success enjoyed by Collen Brothers for most of their tenure. The constructive interaction between Standish and Lyal was crucial to the stability and commercial success of the company in this period.

Collen was undoubtedly well placed to take full advantage of the industrial expansion that began in the early 1960s. The company's reputation for swift and reliable delivery of significant projects certainly benefited Collen in this period. Yet it was not simply the company's undoubted professionalism that facilitated its ability to attract a wide range of clients, but also the network of personal and professional relationships built up by its key members. The company undoubtedly benefited from its long-term connections with the Protestant business establishment in Dublin – Brown & Polson, Fry Cadbury and Jacobs, major clients of Collen Brothers during the 1960s, all had a strong Protestant influence. Often they were Protestant family businesses which shared considerable social and commercial connections with Collen.[51] Yet this was only a single dimension of the firm's success in promoting itself during this period. Lyal Collen in particular developed an extensive web of connections among managers and professionals within the business world in Dublin. Lyal's participation in the Trinity College Dublin Association and Trust, an association representing alumni of Trinity College where he served as chairperson from the 1950s, reflected his genuine interest in education and his commitment to charitable fund-raising to support his alma mater. It is equally evident, however, that his involvement in the Trust helped to consolidate his links with a wide cross-section of TCD graduates, many of them involved in the higher reaches of business in Dublin. More-

---

49. Interview with Paddy Wall, 28 April 2009.

50. Hoffman, 'It took 200 years to build', *Business and Finance*, vol.17, no.33, 30 April 1981.

51. Interview with Paddy Wall, 28 July 2009; Interview with Leo Crehan, 29 June 2009.

over, Lyal Collen emerged as one of the most prominent figures in Irish business in his own right by the early 1970s, serving as a director of a series of major companies, including Cement Roadstone Holdings (CRH), Ergas, Mobil and Celtic Oil.[52] He was appointed in March 1974 as chairman of Bord na Mona, in succession to Aodhogán O'Rahilly, a former revolutionary who had become one of the leading industrialists in post-independence Ireland.[53] Lyal's multiple connections proved an invaluable asset to his own company, doing much to pave the way for Collen's substantial role in delivering major industrial projects.

Standish Collen made a very different but equally central contribution to the development of the business. He was the central figure in the management and organization of the company and all major decisions involving tendering, pricing and the implementation of contracts were referred to him. Standish had a keen appreciation of financial and business constraints, and he was strongly averse to significant risk-taking in business.[54] He had a well-deserved reputation among the company's managers for integrity, shrewd business judgement, meticulous attention to detail and professional conservatism. He gave a revealing insight to his *modus operandi* in an interview with *Business and Finance* towards the end of his long career, when he discussed the advantages of running a family business. He commented that family-owned companies were under 'far less pressure' to deliver the highest possible profit from every transaction and could even forego short-term gain to consolidate a long-term relationship with clients: 'We sometimes make mistakes. When that happens I can say 'Right. Let's pull it down and do it again.' We can do that within the family. I'm not sure that we could do it if we had shareholders demanding profit all the time.'[55]

Standish was hardly indifferent to making a profit, but his comments nonetheless underlined a profound conviction that relationships counted for more in the end than short-term transactions. He was also entirely unwilling to risk the long-term stability of the business or the central place of the family within the enterprise in pursuit of short-term gain. His underlying philosophy

---

52. *Irish Times*, 'Directors of Cement Roadstone announced', 24 October 1970; *Irish Times*, 'New Irish company to seek oil, gas off Irish coast', 13 June 1972.

53. *Irish Times*, 'Chairman of Bord na Mona', 4 March 1974.

54. Interview with Dr Brian Bond, 30 April 2009; Interview with Paddy Wall, 28 April 2009.

55. Ronnie Hoffman, 'It took 200 years to build', *Business and Finance*, vol.17, no.33, 30 April 1981.

was undoubtedly influenced by his personality and his experience as an officer in the Second World War, but it also reflected the traditional business model of Collen Brothers, which was low-key, professional and wary of excessive risk or ostentation. Standish's caution sometimes frustrated his younger managers and he may have left some potentially lucrative business opportunities behind, but his conservatism also guaranteed the stability of the company during an era of expansion.

Standish was deeply engaged in equestrian pursuits throughout his life. He participated in point-to-point racing as a young man and later raised horses on his own land at Streamstown, Malahide.[56] He was a prominent figure within the RDS, serving as an elected member of the Society's committee on agriculture.[57] Standish took the lead in a successful initiative by the RDS to establish a new complex for horse sales at Ballsbridge in the mid-1970s and was involved in bringing the Tattersalls bloodstock agency to Dublin.[58] Standish also had a lifelong passion for hunting, acting as joint master of the Ward Union Hunt for many years.[59] It was his enthusiasm for horseracing and hunting that led to his friendship with Charles Haughey, who became a member of the Ward Union Hunt during Standish's tenure. The two men also became neighbours in Malahide, after Haughey acquired his estate at Kinsealy in the late 1960s. Haughey, then a rising star in Fianna Fáil, would later become the most formidable and controversial politician of his generation, serving as Taoiseach four times between 1979 and 1992. While it was essentially a personal friendship between the two men, the company undertook a significant project for Haughey during the 1970s – the building of a holiday home on Inishvickillane, the island in the Blaskets owned by Haughey since 1974. The house was originally designed by Garry Trimble, a well-known sculptor, but was redesigned by Collen's architects and engineers when the original design proved too ambitious to implement.[60] Standish's son David oversaw the construction of the house, in conjunction with a local builder, between 1976 and 1978: it was the only significant contract undertaken by the

---

56. *Irish Times*, 'Standish Collen: Builder with a passion for all things equestrian', 8 May 2004.

57. *Irish Times*, 'Election of RDS officers', 8 December 1972; *Irish Times*, 'RDS elects officers', 2 December 1983.

58. *Irish Times,* 'RDS bloodstock sales plan backed', 24 June 1974; *Irish Times*, 'Standish Collen: Builder with a passion for all things equestrian', 8 May 2004.

59. *Irish Times,* 'A-Hunting they won't go, say farmers', 22 October 1966.

60. Interview with Pat Sides, 21 July 2009.

company for Haughey.[61] An obituary notice for Standish following his death in 2004 noted his 'deep and enduring friendship' with the former Taoiseach.[62] Their friendship was based above all on a shared interest in horseracing and equestrian pursuits.[63]

FOLLOWING A WELL-WORN PATH

Collen remained a traditional building company throughout the 1950s, relying to a considerable extent on its existing connections and on regular sources of business. The policies of economic protectionism and agricultural self-suffi-ciency were still in force for most of the decade, despite rapidly accumulating evidence of their failure. The economic conditions in which the company operated were dismal and became particularly grim around the middle of the decade, especially following the Suez crisis in 1956. Employment in both industry and agriculture declined throughout the decade, while emigration reached an unprecedented level; over 412,000 people emigrated from Ireland between 1951 and 1961. The Irish economy established a unique benchmark in 1956–7, when Ireland was the only country in Europe to see a fall in the total volume of goods and services consumed.[64] The building industry could not escape the overall economic contraction and suffered a severe slump in the mid to late 1950s.[65]

It was not surprising that the company depended largely on regular clients and well-established sources of business in such unfavourable external condi-tions. The company's involvement in building grain silos reached its peak during the 1950s, not least because government policy continued to promote the expansion of tillage and flour milling. Collen built reinforced concrete silos for Bolands' Mills in Dublin and Ranks in Limerick. The company also provided flat grain stores for various clients, mainly between 1952 and 1956, including the Odlum group in Waterford, McGees of Ardee and Dowleys in Carrick-on-Suir.[66] Odlums emerged as an important long-term client for

61. Interview with David Collen, 10 June 2009.

62. *Sunday Independent*, 'Standish Collen', 9 May 2004.

63. *Ibid.*; *Irish Independent*, 'The trials of CJ Haughey', 25 July 1998.

64. Ferriter, *Transformation of Ireland*, pp.465–6.

65. Lyal Collen, 'Building' in *Careers in Ireland* (Dublin, 1958), p.16.

66. Lyal Collen Papers, Lyal Collen Note, *Grain Silos*, pp.1–2; River House Archive, *Collen Bros. (Dublin) Ltd.: A Selection of Contracts*, p.2.

Collen, offering repeat work over the following two decades. The company built a new flour mill for Odlums in Portarlington in the late 1970s.[67] Collen continued to build grain silos and mills throughout this period, but other elements of the business began to dominate.

Bridge building was another notable feature of Collen Brothers' activity throughout this period. The company benefited from Standish's wartime experience during his service with the Royal Engineers. He had developed a particular expertise in working with bailey bridges, portable structures consisting of a series of prefabricated steel sections in the form of lattices. Shortly after the war, Collen became the agent in the Republic of Ireland for Thomas Storey and Co., an English company based in Stockport, which manufactured and supplied bailey bridges to the British army.[68] Thomas Storey also manufactured pontoon units called Uniflotes, which Collen regularly used for its own marine engineering work. This agency arrangement conferred a considerable advantage on Collen, which became the sole supplier of bailey bridges and Uniflotes in the Irish state. Standish's expertise and wartime connections opened up a valuable revenue stream for the firm. The company installed bailey bridges for most local authorities throughout the country, securing a monthly rent from the relevant council. While the installation of the bridges was a relatively straightforward undertaking, Collen's pivotal position in the Irish market ensured that it also became a valued source of income. Paddy Wall commented that 'it was a bread and butter job, but we got useful revenue from the council every month for them; we just had to inspect them and make sure they were structurally sound.'[69] The company also provided temporary bridging to meet short-term emergencies, such as the collapse of existing bridges due to flooding or other exceptional weather conditions.[70]

One of the most notable bridging projects undertaken by Collen occurred on their own doorstep in East Wall and was the result of a dramatic natural disaster. A heavy storm swept through the country on 8 December 1954, triggering widespread flooding and severely disrupting road and rail communications. The government proclaimed a national emergency, launching

---

67. Interview with Martin Glynn, 16 June 2009; Interview with Pat Sides, 21 July 2009.

68. Collen Minute book, Notes of Directors' meeting, 9 February 1955; Collen Minute book, Minutes, Meeting of Directors, 3 February 1971, p.2.

69. Interview with Paddy Wall, 28 July 2009.

70. Interview with Paddy Wall, 28 July 2009.

Harky Collen, inspecting a newly installed Bailey bridge at Athlone, 1959.
Collen Brothers (Dublin) photo album.

Operation Rescue to contain the impact of the storm and ease the plight
of its victims. The north inner city bore the brunt of the natural disaster in
Dublin, as the Tolka river burst its banks and Fairview Park was completely
submerged.[71] The railway bridge over the Tolka and East Wall road, which
carried the main Dublin-Belfast line of the Great Northern Railway (GNR),
collapsed under the weight of the floods early in the morning of Thursday,
9 December.[72] The north span of the bridge fell into the river, creating a
makeshift dam which diverted the flooded river, so that about a hundred
feet of the road was washed away and the flood swept further into the city.[73]
Collen's headquarters in East Wall was right beside the bridge and the directors immediately offered to build a temporary bailey bridge across the Tolka.[74]
Their offer was rapidly accepted by the GNR – a contract was issued to Collen

---

71. *Irish Times*, 'An Irishman's Diary', 30 November 2009.

72. Lyal Collen Papers, Lyal Collen Note, *Tolka bridge*, p.1.

73. *Ibid.*; *Irish Times*, 'An Irishman's Diary', 30 November 2009.

74. Lyal Collen Papers, Lyal Collen Note, *Tolka bridge*, p.2.

on Friday 10 December to complete the new bridge and the work began on the evening of the same day.[75]

The company provided three-quarters of the material for the temporary bailey bridge, while the remainder was delivered by the Corps of Engineers of the Irish army. The first priority was to clear away the remnants of the old bridge: Collen used CIE railway cranes to shift the ruined steel girders, while two professional divers from Liverpool worked to cut away the remains of the bridge from the river bed when the tide was in. The debris was fully cleared away by the following Monday, 13 December, ending the threat of further flooding due to the blockage in the Tolka. The construction of the new bridge went ahead about seventeen feet inland from the old structure, with the intention of leaving a clear site for building the new bridge.[76] The company was determined to complete the bridge as quickly as possible – considerable urgency was attached to the project because Amiens St station was cut off from the Belfast line as a result of the collapse of the bridge.[77] Collen employed over 200 workers on the site and the *Irish Times* recorded that 'work is going on continuously all day and night'.[78] The work also proceeded six days a week, breaking only for Sundays and for two days on 25 and 26 December.[79] The temporary bridge was completed and placed across the river by 27 December. Following rigorous testing of the new structure, including the running of locomotives across the track, the bridge was opened on 3 January 1955 and a full train service operated from Amiens Street on the next day.[80] The company's action restored the train link between Amiens St station and Belfast with impressive speed and professionalism. The haste with which the project was accomplished did not involve any sacrifice in quality or safety – the new bridge was the first of its kind in Ireland built to carry a load of a hundred tons.[81] Collen's swift and decisive response in an emergency was possible largely because of the company's experience in assembling bailey bridges, and the supply and installation of these bailey bridges remained a valuable element of the company's business up to the 1970s.

---

75. *Irish Times*, 'New bridge for Tolka is full-time job', 17 December 1954.

76. *Ibid.*.

77. *Irish Times*, 'Replacement of Tolka bridge', 23 December 1954.

78. *Irish Times*, 'New bridge for Tolka is full-time job', 17 December 1954.

79. Lyal Collen Papers, Lyal Collen Note, *Tolka bridge*, pp.1–7.

80. Lyal Collen Papers, Lyal Collen Note, *Tolka bridge*, pp.5–7.

81. *Irish Times*, 'New bridge for Tolka is full-time job', 17 December 1954.

Bridging the Tolka; the site on 16 December 1954, while work was in progress on laying the Bailey bridge.

Collen's ability to sustain a long-term connection with clients was exemplified by the firm's enduring association with the RDS. The company built a new library for the Society in 1965, replacing the previous hall which had served a dual function as a library and concert hall. The firm used the existing foundations as well as three walls within the complex to complete a new internal building, thereby reducing the building costs. The new library was a two-storey structure, including a balcony giving access to the second tier of book shelves; it provided space for 60,000 books on display and a reserve book stock of over 200,000 volumes.[82] Collen undertook a much more substantial project for the Society almost a decade later, when the RDS moved to expand its facilities for show jumping and industrial exhibitions on its newly acquired land at Simmonscourt. The company began work early in 1975 on a new pavilion at Simmonscourt, designed by the architect Daithi P. Hanley.[83] The new covered area was a multi-purpose building consisting of an indoor jumping arena, a hall for industrial exhibits, a sales ring for bloodstock and a concert hall.[84] The extension as a whole included four new halls and

82. *Irish Builder*, vol.107, 'New RDS Library', 19 June 1965, p.458.

83. *Irish Times*, 'Way cleared for RDS building plan', 3 January 1975.

84. *Royal Dublin Society*, ed. Clarke and Meenan, pp.73–74; Lyal Collen Papers, Lyal Collen Note, *Royal Dublin Society*, p.2.

covered over three acres: the grand hall, which formed the central section of the pavilion, had a span of 175 feet and was 320 feet long.[85] The new pavilion was completed in ninety days, opening in time for the Spring Show on 5 May 1975. The extension at Simmonscourt provided a spacious and highly flexible covered area, offering considerably greater scope than before for the diverse activities of the Society.

Standish Collen was himself deeply involved in many of the activities of the Society and his equestrian and commercial interests dovetailed towards the end of the decade, when he collaborated with Edward Taylor, then the super-intendent of the RDS, to bring indoor show jumping to Dublin for the first time. Standish and his son David, who became involved in the firm during the early 1970s, took charge of preparations for the project. They produced a jumping surface in the Simmonscourt Pavilion, utilizing the company's Blaw-knox road paver and a specialized mixture of clay/turf mould and salt that would not heave in varying temperatures and was sufficiently firm to allow show jumpers to perform.[86] The Dublin Indoor Show of 1980 was highly successful, attracting top class show jumpers from Britain and continental Europe. The event benefited the RDS, RTÉ – which provided live coverage of the show – and the company itself.[87] Collen played an integral part in the expansion of the RDS, maintaining an enviable record of completing major projects for the Society in every generation since the 1880s.

The firm followed a well-worn path for most of the 1950s, focusing on relatively straightforward building projects and benefiting from the loyalty of regular clients, as it had done during the world economic depression two decades before. Lyal acknowledged the difficult economic conditions facing Collen Brothers Dublin during the first decade following its establishment in an article for *Careers in Ireland*, which was published in 1958. He commented that 'the present year (1958) is hardly the appropriate time to advise a young man to enter any branch of the building industry in Ireland as there has been a very definite slump in the industry…'[88] Yet Lyal took a broadly optimistic tone, describing the considerable opportunities available in different areas of the building and civil engineering business. He emphasized that 'there is and always has been opportunity in the building industry for the bright energetic

---

85. *Irish Times*, 'New Pavilion will be show feature', 6 May 1975.

86. Interview with Paddy Wall, 14 January 2010.

87. *Ibid.*

88. Lyal Collen, 'Building', in *Careers in Ireland* (Dublin, 1958), p.16.

type of person.'[89] The relatively upbeat tone of Lyal's commentary reflected Collen's resilience in weathering successive economic storms over the previous century, not least the grim economic conditions of the 1950s: his optimism was soon justified by the beginning of an upturn in Ireland's economic prospects. The transformation of Irish economic policy would open new horizons for the company, paving the way for an unprecedented expansion of the scope and character of its core business.

INDUSTRIAL EXPANSION

Although there was little overt sign of economic revival towards the end of the 1950s, the publication of *Economic Development*, the seminal policy document composed in November 1958 by T.K. Whitaker and other officials of the Department of Finance, signalled a radical reorientation of economic policy.[90] *Economic Development* advanced a programme for economic expansion based on the encouragement of foreign investment in the Irish economy, the stimulation of export-led growth and a gradual move from protection to free trade. As Gary Murphy points out, this involved 'a dramatic reversal of the rhetoric, and to a large extent the practice of all policy, but especially Fianna Fáil policy, since 1932.'[91] The election of Seán Lemass as Taoiseach in June 1959 dramatically accelerated the pace of policy change: the new Taoiseach made accession to the European Economic Community (EEC) a key objective of his government and took the lead in dismantling the protectionist regime that he had done a great deal to establish in the 1930s. The transformation of Irish economic policy had profound implications for Collen, generating greater industrial development and employment, which in turn meant the initiation of large-scale industrial building projects. The policy changes also opened up new commercial opportunities for the company due to increased foreign investment in Ireland.

The company's engagement with major industrial projects preceded the far-reaching policy changes and the economic expansion of the early 1960s.

89. *Ibid.* pp.16–19.

90. Gary Murphy, *In Search of the Promised Land: The Politics of Post-War Ireland* (Cork, 2009), pp.130–2.

91. Gary Murphy, 'From economic nationalism to European Union', in *The Lemass Era: Politics and Society in the Era of Seán Lemass* (Dublin, 2005), ed. Brian Girvin and Gary Murphy, p.30.

The company built a new factory and offices for the Irish subsidiary of Fry Cadbury, the British chocolate manufacturer owned by Cadbury Brothers, at Coolock in the 1950s. Fry Cadbury had previously been based just beside Collen's headquarters on East Wall Road and the contract involved the relocation of their factory and offices to Coolock at a cost of about £500,000.[92] The new plant, which followed a modern industrial design and layout, was constructed on a greenfield site in Coolock; it was described as 'a factory in a garden' by representatives of Fry Cadbury.[93] Lemass, then still Minister for Industry and Commerce, opened the new factory on 31 May 1957, underlining its status as a major industrial development. Lyal Collen added a theatrical flourish to the formal opening ceremony, presenting Lemass with a golden key to open the factory and a chocolate drinking cup, originally produced in France in 1805.[94] Lyal was instrumental in securing the contract with Fry Cadbury, which was an early instance of his significant role in acquiring new business for the firm.

Fry Cadbury, Coolock; factory and offices built by Collen Brothers during the 1950s. Collen Brothers (Dublin) photo album.

92. Interview with Paddy Wall, 28 April 2009.

93. *Irish Times*, 'New Fry-Cadbury factory opened', 1 June 1957.

94. *Ibid.*

Collen was very successful in winning a substantial share of building contracts associated with the expansion of Ireland's industrial sector between the early 1960s and late 1970s. The company undertook a striking variety of projects, for a mixture of new and traditional clients. In 1960 Collen built a new factory and offices, complete with a shell roof, for Brown & Polson, a British manufacturer best known for producing starch as a food product, who were also a long-standing client of Collen. Ove Arup & Partners acted as the design consultants for the project.[95] Collen also delivered a pipe factory for Asbestos Cement Pipes Ltd. in 1963, working in conjunction with Irish Cement. Another major project was completed by May of the same year, when Lord Boyd opened the new building for the Royal Exchange Assurance group in College Green.[96] Paddy Wall recalled that the project required a particularly deep excavation of the site: '…we had to excavate right down to put in foundations for it, it was quite a high building for its day.'[97] The new building provided offices for the five insurance companies incorporated within the Royal Exchange.

The company was heavily involved in several phases of construction work for another long-standing Dublin business, Roches Stores. The firm provided a new building for the department store in Henry Street during the early 1960s. A decade later Collen resumed work on the same site, undertaking the extension and reconstruction of the store, which was completed in 1976.[98] Much of the company's work in this period consisted of extensive building projects, which were carried out over several years and demanded a large workforce. Collen employed over 150 workers at the Fry-Cadbury site in Coolock, and a similar number worked on the city centre site at Roches Stores.[99] The high number of workers on the two sites was not at all unusual and indicated the company's reliance on directly employed labour rather than subcontractors. The large majority of the firm's workforce at the time consisted of direct employees, reflecting general practice throughout the Irish construction industry in the 1960s as well as the company's preference for using its own employees during this period.

95. *Collen Brothers (Dublin) Ltd. — A Selection of Contracts Since 1960*, p.1.

96. *Irish Times*, 'New assurance offices opened by Lord Boyd', 23 May 1963.

97. Interview with Paddy Wall, 28 April 2009.

98. *Collen Brothers (Dublin) Ltd. — A Selection of Contracts Since 1960*, p.1.

99. Interview with Paddy Wall, 28 April 2009.

CRETESTONE

The expansion of commercial opportunities in the 1960s, combined with Lyal's impressive range of connections, enabled Collen to branch out in new directions: a notable initiative was the acquisition of a dedicated supplier of pre-cast concrete products. The proprietors secured a controlling interest in Cretestone Ltd, based in Ballyfermot. The company was originally founded by Charles Lethbridge in the 1930s and was reconstituted with a wider range of investors in 1944. Robert C. Booth, a prominent member of the Protestant business elite in Dublin, became chairman of the board, which also included other members of the Dublin business community.[100] Lethbridge, however, remained the managing director and was the dominant influence within the company until his retirement in 1955.[101] Collen's association with the company began in March 1956, when Lyal was appointed as a director – he owed his place on the board largely to his connections with the Booth family, who were leading members of the Methodist community in Dublin.[102] Lyal was particularly friendly with Lionel Booth, Robert's nephew, who became a Fianna Fáil candidate and was elected as a TD for Dún Laoghaire-Rathdown in 1957.[103] The proprietors of Collen Brothers took over the ownership of Cretestone in the early 1960s. Lyal and Standish bought out the Booth family and all but two of the original shareholders at a price of £3 per share on 30 October 1961.[104] The new regime maintained a degree of continuity with the past. Henry Lebioda, the principal manager of the Cretestone factory before Collen acquired the company, remained a member of the board until 1975, while John G. Douglas, the heir of one of the founding directors, also held a minority stake in the company and served as a director for four years.[105] Yet while Cretestone remained in theory a separate company, it was firmly controlled by the Collen family directors. Standish and Lyal were the majority shareholders and alternated as chair of the board. Moreover, the vast majority of their shares were transferred to Collen Brothers itself in 1966, making the

---

100. Cretestone (1944) Ltd., List of directors, 5 October 1944.

101. Cretestone Minute Book, Minutes, Annual General Meeting, 15 November 1955.

102. Cretestone, Minutes, Meeting of Directors, 8 March 1956.

103. *Irish Times*, 'Fianna Fáil certain of clear majority: 23 seats undecided', 7 March 1957.

104. Cretestone, Minutes, Meeting of Directors, 24 August 1961; Minutes, Meeting of Directors, 30 October 1961.

105. Cretestone, Minutes, Meeting of Directors, 24 September 1961.

firm the major shareholder in Cretestone.[106]

The large-scale contracts undertaken by Collen Brothers, particularly in Tallaght and Dublin Port, created a considerable demand for pre-cast concrete products. Cretestone therefore became an integral part of Collen's operations over the decade and a half following its acquisition, providing pre-cast concrete products for various projects completed by the firm. Several key managers from Collen Brothers served as directors of Cretestone, including Archie Harris (who resigned his directorship after only seven months citing pressure of work), John Griffin and Stratton Sharpe.[107] The proprietors delegated considerable responsibility within Cretestone to Sharpe, who had overseen the Fry-Cadbury project during the 1950s. Sharpe was heavily involved in running Cretestone on a day-to-day basis; he was appointed as managing director in April 1968 and oversaw its operations for the following three-and-a-half years.[108] Tom Saville, another civil engineer with Collen Brothers who

Fitzwilton House, Dublin: Cretestone wall cladding, 1970s.
Collen Brothers (Dublin) photo album.

106. Cretestone, Minutes, Meeting of Directors, 8 March 1966.

107. Cretestone, Minutes, Meeting of Directors, 12 October 1971; Minutes, Meeting of Directors, 4 September 1973.

108. Cretestone, Minutes, Meeting of Directors, 28 March 1968.

played a significant part in the redevelopment of Dublin port during the 1970s, took over as managing director of Cretestone in October 1971, while Sharpe remained on the board as deputy chairman.[109] Sharpe's association with Cretestone ended in 1974 at the same time as his departure from Collen Brothers.[110] Cretestone was a significant element of the Collen family enterprise from the early 1960s until the mid-1970s, reflecting the overall expansion of the firm during this period.

Yet the acquisition of Cretestone proved a temporary expedient rather than a strategic move to diversify the business. It did not signal the beginning of a long-term venture by Collen in the manufacturing of pre-cast concrete. The family proprietors moved to run down Cretestone's operations during the second half of the 1970s; Lyal and Standish, who held sole control of Cretestone by 1975, decided to realize the value of its assets rather than keeping the company in operation. They sold the factory, land and buildings owned by Cretestone on Killeen Road in Ballyfermot to Readymix on 21 September 1977, receiving £320,000 for their investment.[111] Readymix used the property as a concrete plant to supply its operations in the expanding suburban areas of west Dublin. Cretestone itself remained in existence for several years while the directors settled its outstanding liabilities and reached terms with its debtors, but it was essentially a shell company that did not undertake any further operations. The incorporation of Cretestone within the Collen family enterprise was a pragmatic response to commercial demands; it did not foreshadow the emergence of a multi-unit business enterprise.

PROPERTY DEVELOPMENT – INDUSTRIAL ESTATES

The company significantly extended the scope of its activity in the 1970s, engaging in the development and management of industrial estates for the first time. Collen Brothers had never previously acted as a developer and was very much a conventional building contractor for the first hundred years of its existence. Lyal Collen took the lead in promoting a new departure by the company. He acquired land in Tallaght from Redmond Gallagher, proprietor of Urney

---

109. Cretestone, Minutes, Meeting of Directors, 12 October 1971.

110. Cretestone, Minutes, Meeting of Directors, 31 July 1974; Collen Brothers, Minutes, Meeting of Directors, 31 July 1974.

111. J&E Davy brokers to Cretestone (1944) Ltd., 21 September 1977; Cretestone, Minutes, Meeting of Directors, 21 September 1977.

Chocolates, during the 1960s and this became the basis of a far-reaching new business venture over the following decade.[112] Tallaght experienced a population boom in the 1970s, developing from a village on the outskirts of Dublin into an important satellite town and later a major suburb of the city. The rapid expansion of housing in south-west Dublin created both a ready-made workforce for new industries and a strong demand for jobs among the new suburban communities – planners for Dublin Corporation estimated in 1971 that Tallaght would require about 20,000 jobs in different branches of industry over the following fifteen years.[113] The increasing volume of foreign investment, especially generated by trans-national corporations who were interested in locating in Ireland due to favourable tax incentives and its proximity to the larger European market, also made the development of industrial estates an attractive commercial prospect. The circumstances were very favourable for the company's entry into industrial property development, while the timing and location for its initial venture were well chosen. Collen began to develop an industrial estate on a site of 135 acres in Tallaght in 1966. The project involved the design of the layout to the estate and the provision of the necessary infrastructure, including roads; the main road through the estate was Airton Road, which was the first to be constructed.[114] The company planned to build up the first phase of the estate gradually over a seven-year period.[115]

The directors aimed to reap a double benefit from their initial investment. Collen sold sites to companies for industrial development, while also securing the contracts to build (and usually design) the factories or offices on them for their clients. The company also frequently built advance factories, speculative units developed prior to securing a client, in the new estate.[116] Some of the first occupants were traditional clients of Collen, notably Gallaher's tobacco company, who established a factory and offices at Tallaght in 1968: Collen built the new factory and subsequent extensions to it in 1969–70.[117] Several leading Irish companies acquired sites on the estate, including the Bank of Ireland and Jacob's, the Irish biscuits manufacturer, who moved their factory complex

---

112. Interview with Paddy Wall, 28 April 2009.

113. *Irish Times*, 'Collen Bros provides jobs at Tallaght', 3 November 1971.

114. Correspondence with Dr Brian Bond, January 2010.

115. *Ibid.*

116. Collen Brothers, Minute book, Minutes, Meeting of Directors, 1 September 1971.

117. *Collen Brothers (Dublin) Ltd. – A Selection of Contracts Since 1960*, p.1; *Irish Times*, 'Gallaher plans extension to Tallaght factory', 7 October 1969.

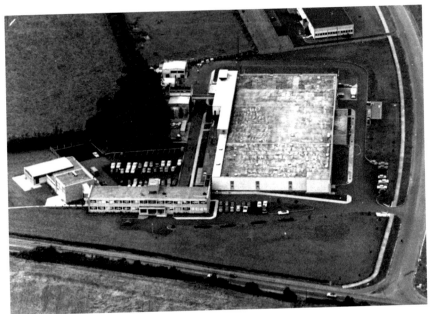

Gallaher's factory, Tallaght industrial estate. Collen Brothers (Dublin) photo album.

from the city centre to Tallaght.[118] But the new industrial estate also attracted an impressive selection of multinational companies seeking to acquire a base in Dublin. The US computer giant, IBM, and Sandoz, a Swiss firm specialising in pharmaceuticals, were among the first wave of companies to establish a presence at Tallaght.[119] American car manufacturer General Motors followed the same path in the early 1970s. The US corporation negotiated an agreement with Collen in May 1973 to locate its new factory for the production of electrical wiring harnesses for Opel and Vauxhall cars at Tallaght.[120] Perhaps not surprisingly, the Irish company secured the lucrative contract to build the factory, which carried a total value of over £1.4 million.[121] The Collen industrial estate continued to develop steadily throughout the 1970s, as the demand for new factories and warehouses remained high up to the end of the decade.[122] Many of the leading brand names of international business were represented at Tallaght, including the music conglomerate Sony and the

---

118. *Irish Times*, 'Jacob city site on the market', 26 January 1972.

119. *Irish Times*, 'Collen Bros provides jobs at Tallaght', 3 November 1971.

120. *Irish Times*, 'General Motors plans to locate plant at Tallaght', 19 May 1973.

121. *Collen Brothers (Dublin) Ltd. – A Selection of Contracts Since 1960*, p.1.

122. *Irish Times*, 'OKM predict that upward trend will continue', 28 March 1979.

international pharmaceutical company, Wellcome.[123] The Tallaght industrial estate became the largest of its kind in Ireland by the 1980s.

Collen's first industrial development at Tallaght was soon followed by a similar venture in Bray. The company began to develop an industrial estate in the early 1970s on land owned by the Earl of Meath; the estate was developed in partnership with the Earl, who was a friend of Lyal Collen.[124] Collen Brothers applied for planning permission to build 'an industrial estate road and ancillary services' along Boghall Road, on the outskirts of Bray, on 21 June 1972.[125] The latest development was much smaller than the Tallaght industrial estate, amounting to about thirty acres. The new estate was one of several established in the Boghall Road development over the following decade by private developers such as Cooney Jennings and by the Industrial Development Authority (IDA).[126] Collen experienced far more intense competition for clients than they had done in Tallaght – the IDA was a particularly formidable competitor, offering various incentives to attract companies to its own nearby estate in Bray.[127] But Collen's venture in the area proved a viable undertaking, at least in part because the company succeeded in attracting new international clients to its estate, including Nixdorf Computers, a leading German information technology corporation. Des Lynch believed that the managing director of Nixdorf was not entirely motivated by financial considerations: 'Mr Nixdorf came over to look at the site and he was quite taken by the leafy trees in Bray and he decided that was the place for him!'[128] Nixdorf opted to establish his factory on Collen's estate despite strong lobbying by the IDA, which sought to persuade him to operate from their development instead. While Collen was successful on that occasion, the episode gave an indication of the intensity of the competition for clients in the new industrial zone. The estate in Bray was never intended to expand on the same scale as its counterpart in Tallaght, and its commercial potential was also constrained by strong competition and by the onset of a severe recession in the early 1980s. Whatever the difficulties, however, it is apparent that on the whole the development of industrial estates

123. Prospectus, *Commercial and Industrial Development by Collen*, 2008.

124. Interview with Des Lynch, 9 June 2009.

125. *Irish Times*, 'Planning Applications', 21 June 1972; *Irish Times*, 'New £600,000 shop/office development for Bray', 1 November 1972.

126. *Irish Times*, 'Considerable activity', 28 January 1981.

127. Interview with Des Lynch, 9 June 2009.

128. *Ibid.*

paid rich dividends for Collen. The development of industrial land not only generated short-term profits especially in the 1970s, but also created extensive business for the company, delivering a steady supply of new contracts for the construction and design of factories and warehouses.

Yet even in the midst of a new and highly successful venture, the company did not abandon its distinctive traditions. The full development of the Tallaght estate took much longer than initially anticipated, due principally to Collen's measured approach to industrial development. Collen from the outset adopted a policy of designing to a high standard, both in the layout of the estate and the quality of the buildings. The roads in the Tallaght industrial estate were noticeably wider than is generally found in such developments, while the buildings were laid out over a wide area.[129] The management of Collen, reflecting their background in construction and engineering rather than property development, placed considerable emphasis on a spacious layout for the industrial estates. More significantly, the directors took a conservative approach in negotiating with clients, which was unusual for developers of industrial property in a period of economic buoyancy. Paddy Wall recalled the concerns of the directors at the time:

> We liked to pride ourselves in thinking that they [the estates] were well laid out, they weren't as dense as the other ones and we picked our clients whereas other people would just have taken them as they came. Now we got a little bit of a reputation for that because we were nearly interviewing people before we allowed them to buy our land and have their factory.[130]

The leading figures in the company were motivated in part by pragmatic financial concerns, seeking clients who could be guaranteed to pay their way: but another factor was also at work, notably Collen's determination to take on reputable clients with an established standing within society. As a long-term company insider commented: 'we were always quite careful that we thought we were taking people who would ... last the pace and were of good standing in terms of their company image and products they produced.'[131] This relatively conservative approach certainly protected Collen's reputation and did not prevent the highly successful development of the industrial estates in Tallaght and to a lesser extent Bray. Yet the company's caution also meant that Collen

---

129. Collen Brothers, Minute book, Minutes, Meeting of Directors, 7 March 1973; Interview with Dr Brian Bond, 30 April 2009.

130. Interview with Paddy Wall, 28 April 2009.

131. Private information.

held on to development land at both estates for a significantly longer period than they had to in purely commercial terms.[132] So the firm did not take full advantage of the buoyant market for industrial and commercial property in the 1970s and was left holding a considerable segment of development land in the much more unfavourable economic circumstances of the following decade.

CIVIL ENGINEERING

Civil engineering was an intermittent feature of Collen's activity since its foundation, but the company's involvement in major civil engineering projects escalated dramatically in the 1960s. This expansion owed a great deal to the initiative of John Griffin and the technical expertise of the company's key staff. As a consequence, Collen undertook a series of large-scale marine engineering contracts, especially in Dublin Port but also extending to Limerick, Shannon and Galway. The volume of trade passing through Dublin Port more than doubled between 1946 and 1972, demanding a far-reaching programme of development to extend the port's capacity. The expansion of the port occurred throughout the post-war period, with initiatives to reclaim land for development, but Collen became a leading participant when a new phase of development was initiated in the mid-1960s. The Dublin Port and Docks Board inaugurated a five-year development plan in 1967, which was designed to double the capacity of the port.[133] Collen secured one of the most important contracts since its foundation in September 1966, when the Board accepted the company's tender for the construction of new quay walls south of the Liffey near Ringsend power station.[134] The value of the contract was £2 million, making it one of the largest in the company's history.[135] It was a bold decision for Collen to tender for such a major civil engineering contract and the move relied entirely on John Griffin's experience. Griffin had worked on quay construction during his early career with Dublin Port and had also gained expertise in marine engineering during his time with the British Admiralty. Griffin undoubtedly played a leading part in persuading Standish to tender for the job, offering reassurance to his boss that the company enjoyed the necessary expertise to undertake

132. Interview with Paddy Wall, 28 April 2009.

133. *Irish Times*, 'Volume of trade has more than doubled in last 25 years', 27 July 1972.

134. Collen Brothers, Minute book, Notes, Meeting of Directors, 12 September 1966.

135. Minutes, Meeting of Directors, 24 November 1966, p.79.

the contract.[136] While the initial contract involved the extension of the South Quays, the company was soon involved in a series of projects on both sides of the Liffey for the Board and other major commercial stakeholders in the port. Collen would play a central role in the expansion of the port and the development of its commercial facilities over the following decade.

The quay walls were constructed using reinforced concrete caissons, following a method that had already been used extensively in the port by the Board's engineers. Caissons are large reinforced concrete boxes that are placed in a line to form the base of a quay wall. In this instance, the caissons were to be floated into position and sunk onto a prepared bed.[137] Each caisson was a huge structure, usually 50 feet long and either 26 or 33 feet wide, depending on the depth of water in which it was to be used: a smaller caisson weighed 1100 tons, while a larger one weighed 1650 tons.[138] While this method of quay construction was far from novel, the company's successful tender relied on an innovative engineering technique for the construction of caissons, developed for the first time by Collen Brothers. The Board had previously built quays in the port, using its own facilities for constructing the caissons, but Collen's engineers pioneered a new system of constructing and placing them. The caissons were built to a height of 15 feet on a raised steel platform suspended from four synchronized winches. This was known as the 'Syncrolift' system, originally manufactured and supplied by a US company – Collen applied the technique to civil engineering construction for the first time.[139] The platform was then lowered into the water and the caissons were floated off to an adjoining area known as the 'building-up berth', where the floating caissons were built up to their final height.[140] Paul O'Sullivan, the chief engineer of the port, acknowledged the innovative nature of the system in a paper to the British Institution of Civil Engineers in 1970: 'in effect a factory type operation has been created with a speedy and regular output of a caisson every nine to ten days.'[141] The development of an assembly line for the production of caissons provided a rapid and efficient method for the completion of quay walls.

---

136. Correspondence with Dr Brian Bond, 30 January 2010.

137. Correspondence with Dr Brian Bond, 30 January 2010.

138. Ibid,; River House Archive, *Development of Dublin Port*, p.1.

139. Correspondence with Dr Brian Bond, 30 January 2010.

140. River House Archive, *Development of Dublin Port*, p.1.

141. P.M O'Sullivan, 'Recent development works in Dublin Port', Paper 7300 S, *Proceedings, The Institution of Civil Engineers, Supplement*, 1970 (vii), pp.181–2.

Caissons under construction, Dublin Port. Collen Brothers (Dublin) photo album.

The preparatory dredging of the river bed was undertaken by separate contractors, the Britannia Dredging Company. The company's engineers levelled a bed of stone by removing excess material in the trench created by the dredging operation to provide an even bed to support each caisson. When the bed was levelled to the required accuracy, each caisson, which following its completion had remained afloat at the building up berth, was towed by a harbour tug to the quay location, moved into position and sunk onto the prepared bed.[142] This founding operation was done on the falling tide, and if the first founding was not sufficiently accurate it was possible to reposition the caisson after low water on the rising tide, several times if necessary. The top of the founded caissons was at half-tide level and this allowed enough time, in the lower half of the tidal cycle, to construct the concrete superstructure.[143] The founding operation proved extraordinarily accurate, so that it was possible to align the caissons to the straight line of the quay wall. Such an accurate alignment of caissons had not previously been regarded as feasible and the works by the firm's engineers were the first to demonstrate that a uniform line of caissons could be achieved.[144] This was a significant engineering achievement by Collen, which resulted in a lower construction cost and a far

142. *Ibid.* pp.82–3; *Development of Dublin Port*, p.1; Interview with Dr Brian Bond, 30 April 2009; Interview with Martin Glynn, 16 June 2009; Correspondence with Dr Brian Bond, 30 January 2010.

143. Correspondence with Dr Brian Bond, 30 January 2010.

144. O'Sullivan, 'Recent development works', *Proceedings, Institution of Civil Engineers, Supplement*, 1970 (vii), p.184.

superior end product.[145] After each caisson was founded and its position checked and approved, it was filled by granular material brought in by tipper trucks. When the caissons were filled, the concrete superstructure was added and quay features such as bollards and ladders were installed. John Griffin was responsible for planning and developing this new system for producing and using caissons to construct a continuous, solid quay wall.[146] Several other civil engineers within the company, including Tom Saville, Martin Glynn, Brian Bond, Pat Sides, Paddy Glanville and Joss Lynam were also deeply involved in Collen's marine engineering operations and played a vital role in delivering successive projects to expand the port.

The construction of the quay walls was a large-scale project that extended over several years. The first caissons for the new South Quays had been founded by 1967, but then the increasing commercial demands on the port generated an unexpected change in the timetable. The completion of the original project was delayed at the instigation of the Board itself, as several leading clients operating from the port made urgent demands for additional facilities. The delay temporarily diverted Collen from its work on the South Quays, but also opened up several beneficial spin-offs for the company, especially in providing commercial facilities for state enterprises based in the port, including the British and Irish Steam Packet Company (B&I) and British Rail. Collen's innovative use of caissons meant that the company was well placed to extend the original contract when other opportunities arose for work in the port. Martin Glynn, who was then the resident engineer for Dublin Port and Docks, commented that 'they had the facility set up and it was very economical'.[147] B&I, which ran a passenger and cargo service from Dublin to Liverpool, announced their intention in March 1967 to introduce roll-on roll-off car ferries for the first time: a car ferry terminal had to be built in Dublin as a matter of urgency to ensure that the new service became operational in May 1968.[148] B&I and the Dublin Port and Docks Board turned to Collen to undertake the project. The company agreed, but insisted that the only way the deadline could be met was by having its staff carry out the design.[149] Collen's original contract with the Board was extended in April 1967 to include the design and construction of

---

145. Correspondence with Dr Brian Bond, 30 January 2010.

146. *Ibid.*

147. Interview with Martin Glynn, 16 June 2009.

148. H.A. Gilligan, *A History of the Port of Dublin* (Dublin, 1988), p.202.

149. Interview with Dr Brian Bond, 30 April 2009.

the car ferry terminal.[150] This was by no means a simple undertaking, as the intended site of the terminal on the north side of the port, near the Eastern Breakwater, was still underwater when the contract was signed and the scheme entailed the reclamation of ten acres of new land.[151]

The B&I Car Ferry Terminal at Dublin Port, completed in May 1968.
Collen Brothers (Dublin) photo album.

Collen's engineers worked closely with technical staff from the Board, including Martin Glynn, who joined Collen shortly after the completion of the project.[152] The reclamation of the land was undertaken by the Britannia Dredging Company; when the site became available in September 1967, Collen moved to provide the new building with impressive speed and efficiency. The car ferry terminal was fully operational by the following May, complete with a new passenger building, jetty, bridge ramp and marshalling areas for cars using the ferry. Standish Collen formally handed over the new terminal to F. Derek Martin, chairman of the Port and Docks Board on 4 May 1968, a few days before the B&I car ferry service was due to begin.[153] The new facility was

---

150. Collen Brothers, Minute book, Notes of meeting, 24 April 1967.

151. Gilligan, *Port of Dublin*, p.202.

152. Interview with Martin Glynn, 16 June 2009.

153. *Irish Times*, 'Port and Docks Board takes charge of new ferry terminal', 4 May 1968.

the first roll-on roll-off car ferry terminal to be opened in Ireland. The delivery of a new commercial building within nine months, on a site just reclaimed from the sea, was no mean feat. It was possible only because Collen was able to take a central role in all stages of the design and construction process. The company was in a position to do this because it had its own design office, and the outcome testified to the advantages of a design and build contract.

Following the opening of the new ferry terminal, the development of new facilities for the unloading and storage of cargo became the immediate priority for the Board and its leading clients. This aspect of the port's building programme was driven by the universal adoption of standard containers to carry a ship's cargo in place of traditional methods of transporting goods in bulk. The move towards 'containerization' was an international reality by the 1960s and significantly influenced the development of Dublin Port.[154] B&I moved to open a new container terminal beside their newly built facility and Collen was awarded the project at the end of 1968. Tom Boyd's notes of the directors' meeting on 9 December recorded 'Design and Construct on the same basis as Car-Ferry Terminal'.[155] The signing of the formal contract, however, was delayed, in part because it involved the negotiation of two separate agreements, one for a new quay and the other for the terminal itself, with the latter not being signed until May 1971.[156] Shortly after the company's initial agreement with B&I, British Rail, which also ran a freight transport business from the port, enlisted Collen's services to construct a similar container terminal in 1969, although on this occasion the rail company itself carried out the design. The British Rail project, which went ahead before its B&I counterpart, involved the construction of a new terminal and quay wall on eight-and-a-half acres of reclaimed land; the container terminal opened for business on the newly extended quay in 1970.[157] The B&I project was a considerably larger development, eventually taking in twenty-eight acres, all of which had to be reclaimed from the sea: it also entailed the construction of a 750-foot-long quay

---

154. Gilligan, *Port of Dublin*, p.202; O'Sullivan, 'Recent development works', *Proceedings, Institution of Civil Engineers, Supplement*, 1970 (vii), p.176.

155. Collen Brothers, Minute book, Notes, Meeting of Directors, 9 December 1968; *Irish Times*, 'Planning Applications', 9 December 1969.

156. Minute book, Notes, Meeting of Directors, 25 August 1969; Notes, Meeting of Directors, 30 April 1971.

157. Collen Brothers, Minute book, Notes, Meeting of Directors, 25 August 1969; Gilligan, *Port of Dublin*, p.203; *Irish Times*, '7.5 million plan by BR gets under way', 23 March 1970.

wall.[158] It was the first major Collen project in which Brian Bond was involved and he recalled the significant challenges presented by working on newly reclaimed land:

> This was all on newly reclaimed land created with dredged fine sand, which I remember on windy days would blow about like a desert sand storm. My first job was to do the setting out to enable the ground surface to be graded to the required levels ... Because of the very large site on which containers were to be stacked, B&I decided that the container handling would be done by large rubber tyre 'straddle carriers' which could go anywhere on the site ... The entire site had to be paved with lean mix concrete and surfaced with asphalt, to carry the heavy wheel loads.[159]

The B&I container and car ferry Terminals at Dublin Port.
Collen Brothers (Dublin) photo album.

The construction of the quay wall began in 1969, but the work on the superstructure of the quay and the building of the new on-shore terminal were held up for several months by a strike at Irish Cement, which cut off the supply of concrete to the construction industry.[160] The project was completed

158. Gilligan, *Port of Dublin*, p.203.

159. Interview with Dr Brian Bond, 30 April 2009.

160. Correspondence with Dr Brian Bond, 30 January 2010.

over a two-year period, with the new container terminal and quay becoming operational in 1972 on an adjoining site to the existing car ferry building.[161] The B&I terminal was undoubtedly a major undertaking, in part due to its location on reclaimed land but also because it was the largest single design and build project undertaken by Collen within the port.

The diversion of Collen's efforts to the north side of the Liffey and the associated developments in the port naturally slowed down the completion of the South Quays. H.A. Gilligan estimated in his official history of Dublin Port that the construction of the South Quays was delayed by approximately four years beyond the original timetable set by the development plan.[162] But there were no complaints from the Board, not least because they had themselves diverted the contractor's primary focus to the north side of the port. The company completed the first section of the medium-depth quays on the south side of the river by November 1969, providing berths with a depth alongside of twenty-five feet at low water and paving the way for a further expansion of the port's capacity.[163] Bristol Seaway, another heavy user of the port, transferred its operations from the Custom House quay to the west end of the new quay, with the intention of securing its own container terminal. Collen was swift to take advantage of the opportunity, winning the contract to build the new Bristol Seaway terminal, which opened in June 1971.[164] Collen managed to combine work on the original contract with other valuable business opportunities associated with the development of the port. The remainder of the medium-berth quay was completed early in 1974 and the new quay became fully operational in the following year with the inauguration of a container service between Dublin and Russian ports.[165]

The firm also provided a new deep water quay on the south wall, intended primarily to rationalize the coal trade into Dublin by providing a single centre for the import and distribution of coal.[166] The project took over five years to complete, in part because it required the reclamation of over thirteen acres of

---

161. Gilligan, *Port of Dublin*, p.203; *Collen Brothers (Dublin) Ltd. — A Selection of Contracts Since 1960*, p.1.

162. Gilligan, *Port of Dublin*, p.203.

163. O'Sullivan, 'Recent development works', *Proceedings, Institution of Civil Engineers, Supplement*, 1970 (vii), p.177.

164. *Collen Brothers (Dublin) Ltd. — A Selection of Contracts Since 1960*, p.1; *Irish Times*, 'Container complex opened by Lenihan', 30 June 1971.

165. Gilligan, *Port of Dublin*, p.204; Interview with Pat Sides, 21 July 2009.

166. Gilligan, *Port of Dublin*, p.204.

land but mainly due to the diversion of the company's energies to the terminals on the north side of the estuary. The new quay, with a depth alongside of
over thirty-seven feet, was completed in 1972, although it only became operational almost two years later due to delays in negotiations between the Board
and the coal importers.[167] The new facility was the first deep-sea quay built for
the Dublin Port authority on the south side of the Liffey.[168]

Work in progress on the South bank quays, Dublin Port.
Collen Brothers (Dublin) photo album.

The completion of the South Quays by the mid-1970s saw the introduction of a new (but short-lived) ferry service between Dublin and Barry, near
Cardiff in south Wales. Sea Speed Ferries, operated by the Morland Naviga-
tion Company, started to operate from Ocean Pier in 1973 and moved to a
new terminal on the South Quays, built by Collen for the Board, in March
1975. But the service rapidly failed to justify its existence and was dropped in
November of the same year.[169]

167. *Ibid.*

168. O'Sullivan, 'Recent development works', *Proceedings, Institution of Civil Engineers, Supplement*, 1970 (vii), p.167.

169. Gilligan, *Port of Dublin*, p.203.

Following the completion of the South Quays, the company acquired several further marine jobs associated with the development of the port, mainly involving the provision of additional facilities for roll-on roll-off ferries. When B&I moved to introduce a new Jetfoil service to Liverpool, Collen designed and built the facility for the high-speed vessels in 1980.[170] It was the first high-speed passenger service operating from Dublin, but it too proved an ill-fated project: as the vessel was too small, it was unable to endure the harsh weather conditions of the Irish Sea and the Jetfoil was soon withdrawn. But a much more significant development undertaken by the firm was the construction of a deep-water quay on the north side of the river in the early 1980s, which was intended to accommodate the larger ro-ro ferries coming into use.[171] Tenders for the new quay were invited on a design and build basis and Collen's tender was based on the use of caissons, which was by then the company's special expertise.[172] The design and construction of the new quay was a major project, overseen on behalf of the company by Brian Bond and valued at about £6 million. The design incorporated an innovative approach to the seaward end of the loading ramp. This comprised a large U-shaped caisson, almost completely submerged, with two towers, one at each end, to support Syncro-lift winches which raised and lowered the ramp to accommodate tidal move-ment. To minimize the time taken to load and unload the ship, the new ramp was unusually wide and had two storeys, enabling vehicles to enter or leave the ferry on two levels.[173] The project, which was completed in 1983, was one of the most impressive and innovative developments undertaken by Collen in the area. It was also the last of the major quay wall projects completed by the firm in Dublin Port.

It is apparent that the company's original job of completing the South Quays proceeded without undue haste, in tandem with a series of associated development projects. Yet the pattern of development undertaken by Collen was certainly in accordance with the objectives of the major client, the Dublin Port and Docks Board; it also reflected the reality that the timing and scope of the port's expansion was dictated as much by commercial demands as official planning. The scale and volume of the projects completed by Collen in the port during this period is striking: the firm was responsible for constructing

---

170. Interview with Dr Brian Bond, 30 April 2009.

171. *Ibid.*

172. Correspondence with Dr Brian Bond, 30 January 2010.

173. *Ibid.*

3500 feet of new quay walls and three container terminals, as well as the two car ferry terminals.[174] The company was the principal contractor in the extension and redevelopment of the port between the mid-1960s and the early 1980s. Collen's expertise in advanced civil engineering techniques, combined with its willingness to undertake design and build projects where possible, exerted a significant influence in facilitating a rapid and ambitious process of expansion. The company's imprint remains visible in much of the contemporary Dublin Port in the twenty-first century.

Aerial view of Dublin Port. Collen Brothers (Dublin) photo album.

Collen installed its own batching plants producing readymix concrete to meet the exceptional demands of the large-scale civil engineering contracts in Dublin. The company set up three plants that operated in conjunction with its own mixer trucks for delivering the concrete to its sites at the industrial estates and Dublin Port.[175] The plants were located at Tallaght, the eastern breakwater on the north side of Dublin Port and the company's caisson-building facility

174. *Collen Brothers (Dublin) Ltd. — A Selection of Contracts Since 1960*, p.1.

175. *Irish Times*, 'Standish Collen: Builder with a passion for all things equestrian', 8 May 2004.

on the south side of the Liffey. It was unusual for a contractor to undertake such a move, as most of the industry relied on specialist suppliers of readymix, such as Roadstone and Readymix Ltd.[176] The establishment of batching plants testified to the company's policy of carrying out as much as possible of the work itself, using its own employees, plant and equipment.[177]

While the development of the port undoubtedly represented a central element of Collen's activity during this period, the company's involvement in marine engineering was by no means restricted to Dublin. Collen installed a new roll-on roll-off cargo facility for Waterford Harbour Commissioners in 1973 – the company provided the components, including a pontoon, and equipped the new facility in the autumn of 1973.[178] More significantly, Collen won a contract with Limerick Harbour Commissioners in 1972 to provide a new jetty at Shannon, which was intended to serve a new terminal for tankers bringing in aviation fuel for Shannon Airport; previously the fuel was landed at Foynes or Limerick city and brought overland to the airport by road tankers.[179] The tanker jetty involved a high level of innovation in marine engineering, not least because it presented a different technical challenge to the construction of quay walls. The client's design of the jetty at Shannon called for three widely separated caissons to act as isolated dolphins, linked by footbridges. This use of caissons as individual dolphins had already been adopted in the B&I car ferry terminal in Dublin, but these were close to the shore. The company's engineers at Shannon prepared and levelled individual stone beds to support three caissons far out in the estuary: the isolated mooring points were equipped with fenders and bollards so that the tankers could be tied up alongside. The new structure was connected to the shore by a number of steel bridges supported by small caissons, allowing a pipeline to pump oil ashore from the tankers.[180] It was a complex and difficult undertaking, which required considerable ingenuity. The caissons were built to the necessary specifications in a dry dock in Limerick and towed over twenty miles down the river to Shannon Airport. It was only possible to tow the caissons down river on top of a falling spring tide, which meant that on arrival at the site the current was much too strong to allow the sinking of the caissons. So the

---

176. Correspondence with Dr Brian Bond, 30 January 2010.

177. *Ibid.*

178. *Irish Times*, '£37,000 port contract for Collen Bros', 4 September 1973.

179. *Collen Brothers (Dublin) Ltd. – A Selection of Contracts Since 1960*, p.1.

180. Correspondence with Dr. Brian Bond, 30 January 2010.

engineers were obliged to scuttle each caisson on the mudflats and wait for about two weeks for more manageable tidal conditions, before sinking them to the required position in the estuary.[181] The scheme went smoothly for the first two caissons but did not work quite as intended for the third one. Brian Bond recalled what happened next:

> Suddenly the mud bank gave way beneath the caisson and it slipped down into deeper water, where it was totally submerged at all stages of the tide and was inaccessible. A salvage operation had to be mounted which was a challenge of unbelievable proportions … we eventually succeeded in getting the caisson afloat, undamaged. We were able to use it in the jetty head; it had a happy ending but was challenging stuff.[182]

The company completed the new jetty without further difficulties, allowing tankers to supply aviation fuel directly to the airport for the first time.

Collen undertook another tanker jetty project for Irish Cement three years later, building two berthing dolphins for an existing terminal on Foynes island serving large tankers supplying the cement factory at Mungret, near Limerick.[183] The site was in the middle of the Shannon estuary near Foynes Island, requiring all the work to be done from floating plant. On this occasion caissons were not involved: each of the new dolphins was supported on fifteen large tubular steel piles. The pile driving itself was subcontracted to a specialist German firm from Hamburg, who provided the floating piling rig to do the job.[184] The project passed off rapidly and without incident, not least due to very favourable weather conditions in the summer of 1975. Brian Bond, who was site agent for the project, found his work being admired by a travelling school of dolphins:

> In a marine job like that out in the middle of the Shannon estuary, there could be down time, possibly for days on end, if the weather was against you. But it was a superb summer, one of the best summers ever, we had dolphins frolicking around us and sunny days, it was just magic, and I think the most profitable job I was ever involved with, because the weather was so much with us.[185]

While neither project was on the same scale as the development of Dublin Port, the construction of the tanker terminals testified to the advanced technical expertise enjoyed by Collen during the 1970s.

---

181. Interview with Martin Glynn, 16 June 2009; Correspondence with Dr Brian Bond, 30 January 2010.

182. Correspondence with Dr Brian Bond, 30 January 2010.

183. *Collen Brothers (Dublin) Ltd. – A Selection of Contracts Since 1960*, p.1.

184. Correspondence with Dr Brian Bond, 30 January 2010.

185. Interview with Dr Brian Bond, 30 April 2009.

Mooring dolphins at Foynes. Collen Brothers (Dublin) photo album.

The company's impressive record in marine engineering opened the way for its participation in a more important and commercially valuable project on the Shannon estuary during the late 1970s. Alumina Contractors Ltd, an international syndicate led by a Canadian company, Alcan, moved to develop a new alumina plant at Aughinish island in 1978. The factory was designed to process bauxite so as to produce 800,000 metric tons of alumina annually. It was the largest single capital investment project undertaken in the state up to that time.[186] The new development required a deep-water marine terminal on the island to handle the large bulk carriers importing bauxite. Collen joined forces with Christiani & Nielsen, a UK subsidiary of a Danish civil engineering company, to tender for the contract.[187] Christiani & Nielsen was an international corporation specializing in bridges, marine works and other reinforced concrete structures. The Irish company lacked the specialized plant to undertake the project alone, while Christiani & Nielsen was

186. *Irish Times*, '£280m Alcan plant to be built on Shannon site', 24 November 1977; *Irish Times*, 'Agreement signed on $500 million Alcan plant', 28 January 1978.

187. *Irish Times*, 'Dublin, London firms get £10 million Aughinish contract', 18 January 1979.

seeking an Irish partner with local knowledge and an established record in civil engineering. They succeeded in winning the contract, valued at £10 million, with Alumina Contractors in January 1979.[188] Their partnership at Aughinish was known as Christiani & Nielsen-Collen, reflecting the Danish company's status as the more substantial partner with a majority stake in the joint venture, which ultimately proved to be a financial success for both parties, although at times this seemed to be an unlikely prospect as the project was far from trouble free.[189]

The plans for the marine terminal involved the construction of a jetty to receive large cargo vessels, which would carry raw materials to the factory, smaller ships for the export of alumina from the island and tankers for importing fuel and other liquids for processing purposes.[190] Brian Bond acted as the site director for Collen's side of the project, with Bill Fleeton as the company's senior representative on site. The Irish company provided several key management staff, while their Anglo-Danish counterparts supplied the project manager, technical staff and almost all of the necessary plant. The works were designed by Rendell Palmer and Tritton, a well known firm of consulting engineers based in London.[191] The first requirement was to establish the foundations for the jetty: vertical tubular steel piles of two metres in diameter were driven into the river bed to support the jetty head.[192] Brian Bond acknowledged the importance of Christiani's participation in providing the equipment and expertise to complete the project: 'Christiani's had the gear to do this … You're talking about big plant to handle this, which nobody in Ireland would own.'[193] The jetty head itself was formed of a pre-cast concrete deck, which was 285 metres long and about 40 metres wide; it carried mechanical loaders for the transport of ore and alumina, as well as pipelines to pump liquid products being imported to the terminal.[194] In effect a pre-cast concrete factory was created on the site, an area of particular expertise for Collen, whose engineers had worked with pre-cast concrete since the 1940s and had recently completed similar projects in the same region.

---

188. *Irish Times*, 'Dublin, London firms get £10 million Aughinish contract', 18 January 1979.

189. Correspondence with Dr Brian Bond, 30 January 2010.

190. *Ibid.*

191. *Ibid.*

192. *Ibid.*

193. Interview with Dr Brian Bond, 30 April 2009.

194. *Irish Times*, 'Dublin, London firms get £10 million Aughinish contract', 18 January 1979.

While the partners collaborated effectively in meeting the technical challenges at Aughinish, the project did not proceed as smoothly as they had hoped.
Collen rarely experienced industrial disputes in this period and had no strikes
initiated by its own employees, but the joint venture could not escape industrial relations strife at Aughinish. The umbrella company employed subcontractors on a cost-reimbursable basis, adopting a policy that was designed
to enforce its control of labour relations on the site. It was an exceptional
practice for a client to reimburse the costs of contractors, as most contracts
in Ireland were conducted on the basis of a fixed price contract, which meant
that remuneration and industrial relations became the sole responsibility of
the contractor. But the adoption of a reimbursable contract gave the client
a continuing financial interest in the day-to-day details of labour costs and
required its approval of any bonus payments or plus rates which its contractors
wished to pay.[195] Alumina Contractors attempted to assert tight control over
pay rates and industrial relations on the site, restricting the normal role of the
individual contractors and of employer organisations such as the Construction Industry Federation (CIF).[196] It was hardly surprising that Alumina
Contractors soon became embroiled in a series of disputes over pay rates
with workers employed by its subcontractors, which significantly delayed the
completion of the project. An unofficial strike by twelve carpenters in June
1979 brought work on the marine terminal to a halt for five days; although
only a small number of workers were directly involved, the remainder of the
650 employees on the site refused to pass the pickets.[197] This was only one
of a series of strikes, either by workers at the marine terminal itself or more
frequently on the larger land-based site.[198] Aughinish soon became a byword
for severe industrial relations strife.

The Collen representatives were deeply dissatisfied with the management
of industrial relations on the site by Alumina Contractors. The joint venture
partners attempted to take remedial action of their own in the autumn of 1979.
Christiani & Nielsen-Collen advertised in national newspapers for a 'senior
Industrial Relations Officer' in August 1979, who would take responsibility
for 'all aspects of industrial relations on the contract…'[199] This move by the

195. *Irish Times*, 'Aughinish report hits at industrial relations', 27 July 1981.

196. Interview with Dr Brian Bond, 30 April 2009.

197. *Irish Times*, 'Strike which halted plant project ends', 20 June 1979.

198. *Irish Times*, 'Alumina strike still on', 10 October 1979.

199. *Irish Times*, 'Industrial relations, civil engineering', 16 August 1979.

partners testified to their exasperation with the client, but could not resolve the wider labour relations issues at Aughinish, nor did it prevent strikes by employees of other contractors. The island site shut down completely in April 1980 following a major unofficial action over bonus payments by almost 200 workers employed by Wimpey-Hegarty, the main contractor for the factory itself. Alumina Contractors declared that the site had become 'unmanageable' as a result of sustained unofficial action and promptly closed the entire plant.[200] This dispute brought the project to the brink of collapse, as the client's managers indicated that the owners might cut their losses and abandon the enterprise. Following prolonged negotiations between the owners and officials representing the group of unions, work on the plant eventually resumed almost two months later.[201] It was little consolation to Christiani & Nielsen-Collen that the dispute had nothing to do with its own industrial relations practices or indeed that one of the shop stewards involved in organizing the unofficial strike declared on 30 April that 'one of the subcontractors, a Danish-Irish company, had an excellent scheme, giving workers from £78 to £90 a week bonus'.[202] The ineffective management and unofficial militancy which characterized the plant at Aughinish was vividly highlighted in a report commissioned by the government in 1980. The report, issued by a committee of three experienced industrial relations practitioners in May 1981, vindicated Collen's reservations about the management of the site, concluding that 'the overall strategic planning by the client for industrial relations on the site was seriously deficient.'[203]

Despite the significant delay imposed by successive industrial disputes, the Christiani & Nielsen-Collen partnership itself worked well and the new marine terminal became operational in 1982.[204] The new jetty head provided two berths for incoming and outgoing cargo traffic: the outer berth was designed to receive vessels of up to 70,000 tons deadweight, which would carry bauxite to the island and export alumina from it.[205] The second berth was intended for smaller vessels importing fuel oils for steam generation and

---

200. *Irish Times*, 'Unofficial strike halts work at Alcan plant site', 30 April 1980; *Irish Times*, 'Terms for return at Alcan plant worked out', 13 June 1980.

201. *Irish Times*, 'Alcan men return today to their £215 a week jobs', 25 June 1980.

202. *Irish Times*, 'Unofficial strike halts work at Alcan plant site', 30 April 1980.

203. *Irish Times*, 'Aughinish report hits at industrial relations', 27 July 1981.

204. *Irish Times*, 'Massive Shannon alumina plant nears completion', 29 October 1982.

205. *Irish Times*, 'Dublin, London firms get £10 million Aughinish contract', 18 January 1979.

Tanker jetty at marine terminal, Alumina plant on Aughinish Island.
Collen Brothers (Dublin) photo album

liquid caustic soda for processing in the new works. The newly built jetty was connected to the island by an exceptionally long approach way of 850 metres.[206] The completion of the jetty at Aughinish was a considerable technical achievement, which eventually turned out to be a financial success despite the formidable industrial relations obstacles.

The joint venture at Aughinish remained, however, very much an exceptional initiative by Collen Brothers. The firm's directors did not authorize any further joint projects with international partners, and although there were discussions with Christiani & Nielsen about further collaboration in Britain, they came to nothing. Standish's caution was undoubtedly a significant factor in ruling out further ventures: he was certainly reluctant to engage in international ventures that would be more detached from the influence of family members. It is apparent that the dominant shareholders, in particular Standish, had no great desire to break into international markets and certainly no inclination to turn the company into a multinational corporation. The company's joint venture with Christiani & Nielsen was an impressive but solitary landmark rather than a precedent for further international collaboration or the beginning of a distinctive new path for Collen Brothers.

206. *Ibid.*

LIGHTHOUSES

Collen Brothers' international reputation in advanced civil engineering methods proved an invaluable asset when the firm undertook two rare but successful ventures in lighthouse construction. Collen Brothers built their first lighthouse in over fifty years during the late 1950s. The firm won a contract with the Commissioners of Irish Lights to replace a lighthouse originally built in the early nineteenth century on Inishtrahull island, six miles north of Malin Head in Co. Donegal. The company completed a modern lighthouse in 1958: the new tower, which was composed of reinforced concrete, was built on the highest point of the island, 133 feet above sea level.[207] The firm's engineers faced a much more demanding test of their technical ingenuity a decade later, when Collen undertook the construction of a new lighthouse in Galway Bay. The contract, which was awarded by Galway Harbour Commissioners in 1969, required the establishment of a new man-made structure in some twenty-five feet of water off Mutton island.[208] Martin Glynn, a native of Galway, took charge of the project; he recalled it many years later as a 'very interesting

Inishtrahull Lighthouse, 1958. Collen Brothers (Dublin) photo album

207. River House Archive, Lyal Collen Note, *Inishtrahull Lighthouse*.

208. River House Archive, Lyal Collen Note, *Galway Lighthouse*.

contract', which was a pardonable understatement of the considerable diffi-culties involved.[209] The company's engineers constructed a circular concrete caisson as the base of the lighthouse – both the base and tower were built ashore in Galway harbour. Then the structure, which was seventy feet high, was floated out to sea and towed to its final location in the bay, where the base section was sunk onto the seabed, resting on a levelled bed of broken stone.[210]

Lighthouse in Galway Bay, 1969. Collen Brothers (Dublin) photo album.

---

209. Interview with Martin Glynn, 16 June 2009.

210. River House Archive, Lyal Collen Note, *Galway Lighthouse*, Correspondence with Dr Brian Bond, January 2010.

The structure was protected against erosion by dumping a substantial quantity of rock around the caisson. The new lighthouse, which was completed
in seven months, represented a considerable feat of marine engineering. The
project gave another demonstration of the firm's expertise in constructing
caissons, but did not serve as the model for future lighthouse construction
on its part. Indeed the Galway lighthouse was the final one built by Collen.
The relatively low financial rewards and high risk involved in such projects
may have discouraged the firm from pursuing further ventures in the area.
While the Galway lighthouse in particular was an impressive achievement,
neither project was very profitable. Collen's successful but brief venture into
the building of lighthouses was not an important or enduring element of
the business, but testified to the wide-ranging portfolio of civil engineering
projects undertaken by the firm and to the ability of its engineers to adapt
successfully to a variety of technical challenges.

EXPANSION AND STABILITY

Collen Brothers experienced an uneven but generally impressive upward
trajectory following the economic stagnation of the mid-1950s. The company
expanded dramatically for over two decades, resulting in significant increases
in the extent of its workforce and the value of its business. The firm's payroll
testified to the most sustained period of expansion since its foundation. Collen
Brothers had 272 staff on its books in 1961: staff numbers increased moderately
to 330 in 1966 and much more dramatically to over 500 by 1981.[211] The company's turnover also showed a gradual increase during this period. The turnover
for the Dublin branch of the firm in the final year before the establishment
of the independent company was £186,817.[212] Collen Brothers, Dublin was
achieving an annual turnover of about £12 million by 1981, an impressive
advance even taking into account considerable rates of inflation during the
intervening period.[213] It was apparent that the company had achieved notable
commercial success in the generation since it became an independent firm.

---

211. Collen Minute book, Note on payroll 1961–66; Hoffman, 'It took 200 years to build',
Business and Finance, vol.17, no.33, 30 April 1981.

212. Collen Papers Portadown, *Dublin Contracts Account for Year ended 31st December 1949.*

213. Hoffman, 'It took 200 years to build', *Business and Finance*, vol.17, no.33, 30 April
1981.

The company greatly extended the scope and character of its activity between the late 1950s and the late 1970s. The firm's entry into property development and management for the first time marked an important and enduring expansion of its core business. Yet Collen did not become a conventional property developer – the firm's priorities in managing industrial estates included the initiation of new construction opportunities and the achievement of high design standards, as well as the obvious objective to make a profit on their investment. The company enjoyed considerable success in delivering major civil engineering projects and was consistently innovative in employing new civil engineering techniques throughout this period. There is no doubt that Collen enjoyed a golden era in marine engineering between the mid-1960s and the early 1980s: it would become apparent in retrospect that it was the peak of a boom in marine engineering construction in Ireland and the high point of the company's involvement in large-scale civil engineering projects. The extraordinary range of the projects undertaken, the impressive commitment of key staff to design and technical innovation and the gradual evolution of its internal management structures all provide persuasive evidence that Collen Brothers was a very different company by 1980 than it had been at the beginning of its independent existence in 1950.

Yet while the changes that occurred over a generation should not be minimized, Collen Brothers remained at its core a family institution rather than a managerial corporation. The firm developed as a successful medium-sized Irish company that showed no desire to grow into a multinational conglomerate, as some major Irish enterprises did in this period. The proprietors were certainly more willing to delegate to trusted employees than before and more inclined to explore new commercial opportunities, but their general approach remained cautious, measured, protective of the firm's reputation and wary of excessive risk-taking. This approach could easily be attributed to Standish's personal philosophy and he undoubtedly exerted a profound influence on the development of the company. But the business model followed by both leading proprietors, Lyal as well as Standish, also reflected long-term core values which formed an integral part of the traditions of the firm.

FIVE

Recession and recovery

The company's fortunes changed abruptly in the early 1980s, as corporate expansion gave way to a sharp retrenchment. Collen Brothers was severely affected by external factors, suffering from the impact of an international economic recession, whose effects were intensified in Ireland by restrictive fiscal measures to control a spiralling public debt. The harsh economic environment led to a significant contraction in the business and to a far-reaching process of internal rationalization by the middle of the decade. The company was obliged to adapt to the loss or decline of key sources of turnover that had been its mainstay over the previous generation. The evolution of the business was influenced too by long-term trends in the construction industry, notably the gradual replacement of direct labour by subcontracting and the increasing importance of information technology. This period also saw a prolonged transition within the company itself, as the dominant figures of the previous generation made way for their successors.

The most significant external factor affecting the company's performance in the 1980s was the rapid decline of the Irish economy. The second oil crisis in 1979 triggered an international economic downturn, which became a full-scale recession in the early 1980s. The impact of the downturn on the Irish economy was exacerbated by disastrous policy errors on the part of political elites, notably excessive government spending in the late 1970s, creating a high level of public debt. The faltering efforts of successive governments to tackle the debt crisis up to 1987 led to a contraction in public spending and the imposition of tax increases, which intensified the severity of the recession. Collen suffered badly from the economic slump, which sharply reduced opportunities in its traditional building domain, while also curtailing new areas of business built up over the previous generation. The demand for industrial property remained buoyant as late as 1979: an analyst from Osborne King and Megran in March 1979 commented on an 'exceptional' level of demand for new factories and warehouses, expressing optimism that the property market would maintain its upward trend in the short-term.[1] Such confidence appeared rational at the time, but it proved sadly misplaced in the harsh economic climate of the early 1980s. The market for industrial property suffered an abrupt decline, as demand for industrial units fell off rapidly due to the recession; there was an oversupply of manufacturing and warehouse space on a national basis by the autumn of 1982.[2] The downturn in the property market severely affected the prospects for development of the company's industrial estates in Tallaght and Bray. Collen had recently embarked on a third phase of development in Tallaght, with Allied Irish Banks (AIB) taking a stake in the venture as a minority partner. The company faced a steep uphill climb in attracting clients for the latest phase of the estate. Paddy Wall recalled that the firm had difficulty either in selling the land or even holding onto it in the hope of an upturn, not least due to pressure from its partners: 'We had terrible trouble trying to sell the land and build factories for people. And when the banks decided that this was going nowhere and they wanted to get out of the land they put pressure on us to try and get rid of the land at whatever cost.'[3] The final sections of the land at Tallaght were sold off at a much reduced value and the company was obliged to accept unfavourable terms to attract clients, particularly towards the end of the process. Irish Tea Merchants purchased one of the final lots on the estate for a new warehouse, but insisted on selecting its own building contractor for the unit in 1983 – it was the only element of the development not constructed by Collen.[4] The final phase of development at Tallaght was painfully slow and produced minimal financial return for the firm.

---

1. *Irish Times*, 'OKM predict that upward trend will continue', 28 March 1979.
2. *Irish Times*, 'Recession slows demand for space', 3 September 1982.
3. Interview with Paddy Wall, 28 April 2009.
4. *Irish Times*, photograph, 19 January 1983.

Despite the increasingly harsh economic environment, the company maintained its involvement in major civil engineering projects during the early 1980s. This work was certainly valuable in the short-term, but offered few grounds for optimism about the long-term continuation of such an important strand of the business. The vast majority of the projects involved the completion of existing marine engineering work, initiated during the heyday of port and harbour development in the previous decade. The most notable – and difficult - project undertaken by the company around this time was an unusually complex civil engineering job. Collen Brothers won a contract with Dublin Corporation in 1976 to construct a new lift pumping station at Ringsend, which was intended to replace the old pumping station commissioned at the same location in 1906. The new station was designed to meet the sewerage requirements of the new Greater Dublin Drainage Scheme, which involved the installation of sewerage facilities to serve newly developed suburban centres including Blanchardstown, Clondalkin and Tallaght.[5] The new pumping station would receive the sewerage from several sources and raise it to a collection chamber at ground level from which it would flow by gravity to new treatment works at Pigeon House on the Poolbeg peninsula.[6] This project extended well into the 1980s, not least because it proved to be one of the most elaborate and difficult civil engineering jobs undertaken by the firm. The contract provided a revealing case study in the pitfalls sometimes involved in a major civil engineering project. The work on the new pumping station began in October 1976, with Martin Glynn acting as the site agent for the company. Brian Bond oversaw the technical aspects of the contract, with responsibility for designing the extensive temporary works. He described the project as 'a colossal technical challenge, it went as wrong as it could. We faced unforeseeable and very difficult ground conditions.'[7]

The job ran into trouble almost from the outset. The first phase of the project was the construction of the circular pump house, which was founded fifteen metres below ground level. The pumping station was built on an area of old reclaimed land, overlying various layers of sand, gravel and silty clay. The site investigation report provided by the Corporation's engineering advisers gave inaccurate information about the ground conditions, indicating that the

5. Dublin Corporation, *The Greater Dublin Drainage Scheme* (Dublin, 1986), pp.1–8.

6. River House Archive, John Taylor & Sons Consulting Engineers, Corporation of Dublin, *Greater Dublin Drainage Scheme: Main Lift Pumping Station – Ringsend*, Contract No.12, vol.4, October 1976.

7. Interview with Dr Brian Bond, 30 April 2009.

silty clay, in which the pumping station was to be founded, was impermeable and would facilitate the construction of a watertight structure.[8] The company's engineers planned the project accordingly, initiating the excavation of the site and the construction of temporary works based on their understanding of the ground conditions. Brian Bond designed a circular cofferdam, a temporary sheet-piled structure required to support the sides of the excavation and establish a safe and dry space to allow the construction of the pump house. The cofferdam was formed by driving sixteen-and-a-half-metre long sheet piles into the silty clay and as the soil was excavated from within it, three reinforced concrete ring beams were constructed to provide support to the sheet piles.[9]

But a 'blow', a sudden inflow of water from the bottom of the hole, occurred in April 1977, threatening to cause the collapse of the temporary works. Although the rate of inflow was relatively small, this was an ominous development as the bottom of the hole should have been completely dry. The inflow appeared to be a piping failure, which could potentially lead to the collapse of the cofferdam.[10] Brian Bond recalled 'it was panic stations, we all wondered what we would do next.'[11] The efforts of the Collen staff on the site managed to reduce the inflow of water but could not stop it completely, and as it was clearly unsafe to proceed, the work was halted and the cofferdam was allowed to flood.[12] The piping failure was a localized outbreak at the south side of the cofferdam and did not threaten any existing buildings. But the company faced an unexpected dilemma in dealing with the instability in ground conditions. Having sought external advice from a British geotechnical engineering firm, Soil Mechanics Limited, Collen's engineers accepted their recommendation to sink twenty-five small relief wells, which extended almost twenty metres underground; this was to relieve the pressure in the ground water beneath the cofferdam.[13] The engineers also acted to extend the sheet piles by three-and-a-

8. A.L Little, B.L. Bond, and W.J. Marshall, *Groundwater Control in the construction of a Dublin pumping station*, *Proceedings of the Ninth European Conference on Soil Mechanics and Foundation Engineering: Groundwater Effects in Geotechnical Engineering*, 31 August–3 September 1987, ed. E.T Hanrahan, T.L.L Orr and T.F. Widdis, pp.183–8.

9. Correspondence with Dr Brian Bond, 11 January 2010.

10. Correspondence with Dr Brian Bond, 5 February 2010.

11. Interview with Dr Brian Bond, 30 April 2009.

12. Dr Brian Bond, *Case History: Main Lift Pumping Station at Ringsend*, 9 March 2000.

13. Little, Bond, and Marshall, *Groundwater Control in the construction of a Dublin pumping station*, *Proceedings of the Ninth European Conference on Soil Mechanics and Foundation Engineering*, 31 August–3 September 1987, ed. Hanrahan, Orr and Widdis, p.184.

half metres, driving the whole cofferdam by the same depth further into the ground. When these additional works were carried out, the excavation of the cofferdam was resumed early in 1978. This solution appeared to work initially and the excavation reached some fifteen metres below ground level.

Ringsend pumping station – cofferdam. Collen Brothers (Dublin) photo album.

But the project suffered a second and much more dramatic setback on 23 May 1978: a second blow erupted at the north side of the cofferdam, almost opposite the site of the first inflow over a year before. This time the rate of flow was much larger and it carried a torrent of silt and sand into the excavation.[14] This was a much more dangerous accident, as it also caused the ground outside the temporary works to subside, threatening to wreck the existing pumping station. Brian Bond, who was in charge on the site when the accident occurred, retained a vivid memory of the collapse over thirty years later:

14. *Ibid.* p.185.

I will never forget that evening. It was half past five in the evening when this happened and the only way to stabilize it was to fill the hole with water to balance the water pressure and so stop the inflow, to stabilize the flow. It was clear that water was flowing under the piles, bringing soil with it and the ground on the outside, you could see it subsiding, so the ground was actually collapsing beside the existing pumping station, a really frightening situation…[15]

The engineers faced the appalling prospect that the old pumping station and sewer would collapse into the ground, putting the existing sewerage system for most of Dublin out of action. It was immediately apparent that only the fire brigade would have the necessary resources to prevent disaster, by pumping sufficient water from the Liffey to flood the excavated hole and discharging hydrants into it. Brian Bond telephoned the fire brigade, only to find that they were in the midst of industrial action:

So I rang the fire brigade and I was told they were on strike and they were only coming if it was a real emergency. So I persuaded them that indeed it was an emergency, so they said 'ok we'll be there'. They turned up and went into action with fire engines and hoses and they pumped from the Liffey and we filled it up safely. But they turned up with the press corps, somebody got a tip-off, so the next thing there was a reporter wanting to get the story and I said 'there isn't a story, you know, this isn't a problem really'. So what happens next the shop steward is in the office, saying 'I thought you said this was an emergency!' The engineer from the Corporation's consultants was on television that night; we couldn't keep the story out of the papers.[16]

While Brian Bond may have failed to downplay the story, his urgent appeal to the fire brigade achieved its primary purpose. The flooding of the hole stabilized the ground and stopped the alarming subsidence next to the old pumping station.[17] But although the immediate danger had passed, the initial temporary works were still incomplete and the formidable technical challenge of completing the underground base for the pumping station remained.

Collen's engineers sought advice from a different expert, Alan Little, an internationally known geotechnical engineer who specialized in earth dams and ground-water control. He recommended forming a slurry trench outside the cofferdam, taken down into the bedrock, some thirty metres below ground.[18]

---

15. Interview with Dr Brian Bond, 30 April 2009.

16. *Ibid*.

17. Little, Bond, and Marshall, *Groundwater Control in the construction of a Dublin pumping station*, *Proceedings of the Ninth European Conference on Soil Mechanics and Foundation Engineering*, 31 August–3 September 1987, ed. Hanrahan, Orr and Widdis, p.185.

18. Correspondence with Dr Brian Bond, 5 February 2010.

The completion of the works required further international expertise, and a French subcontractor, S.F. Bachy, was brought in to construct the self-setting slurry trench.[19] The company's engineers also strengthened the cofferdam by installing additional reinforced concrete ring beams on top of the existing beams, which had been deformed by the accident. The third attempt proved to be the most successful and the excavation was completed without further complications. An impressive range of remedial measures had been required to establish a stable and watertight cofferdam, allowing the construction of the circular pump house itself to proceed.[20] The concrete base slab for the pump house, which was cast as a large single pour of over 800 m³, was finished in April 1980, followed over the next year by the walls and the ground floor slab that supported the pump motors.[21]

The remainder of the project presented further technical challenges but was not attended by the same level of excitement – or trauma – as the construction of the pump house. The engineers designed and supervised the construction of a new screen house and adjoining collection chamber that received the waste from the sewers and transferred it to the pump house itself. There were still serious engineering problems to be solved but they were dwarfed by what it had taken to build the pump house.[22] The engineers benefited from their newfound experience with the ground conditions to develop a workable design for the screen house. A short reinforced concrete culvert, linking the screen house to the pump house, was completed successfully in February 1982.[23] The final phase of the works was the awkward and unsanitary job of diverting the main city sewers to the new station. The original time-scale for the job had been fifty months, but the unexpected scale and complexity of the project ensured that work on the pumping station continued for almost nine years, with the new plant only becoming operational in July 1985.

---

19. Little, Bond, and Marshall, *Groundwater Control in the construction of a Dublin pumping station*, *Proceedings of the Ninth European Conference on Soil Mechanics and Foundation Engineering*, 31 August–3 September 1987, ed. Hanrahan, Orr and Widdis, p.186.

20. *Ibid.* p.188.

21. B.L. Bond, *Upheaval Pressure and Base Failure*, contribution to discussion session, *Proceedings of the Ninth European Conference on Soil Mechanics and Foundation Engineering*, 31 August–3 September 1987, ed. Hanrahan, Orr and Widdis.

22. Correspondence with Dr Brian Bond, 5 February 2010.

23. Dr Brian Bond, *Case History: Main Lift Pumping Station at Ringsend*, 9 March 2000; Correspondence with Dr Brian Bond, 11 January 2010.

It was apparent that the original site investigation report supplied to the tendering contractors had failed to assess the ground conditions accurately. This blunder had serious implications for Collen, who faced a significant delay and a major escalation in costs to complete the project. Following the spectacular accident in May 1978, Brian Bond made a strong case to the Corporation's consulting engineers, John Taylor & Sons, that the problems were due to unforeseeable physical conditions, which were the employer's responsibility under the terms of the contract.[24] While negotiations dragged on for some time, the Corporation's engineers ultimately accepted the company's case and Collen was reimbursed for the extensive additional costs that were incurred in completing the job.[25] The company would have had a serious financial problem if matters had turned out differently. Martin Glynn was emphatic about the importance of the successful contractual negotiations for the firm: 'that would have been a major problem if we hadn't got that. It would have been financially very difficult for the firm at the time. But the Corporation were fair enough, the rules were there, they played by the rules … and people were treated fairly.'[26] The extra costs of the civil works more than trebled the original tender price: the contract was initially valued at £1.6 million in 1976, but the final account amounted to £4.5 million eight years later. The company was fortunate that the accident occurred at a relatively early stage of the contract, at a time when the source of the problem was readily identifiable and well before its own civil engineering operations reached an advanced stage.

The completion of the new main lift pumping station allowed the full implementation of the Greater Dublin drainage scheme. The new station received most of the sewerage from both new and old urban schemes, deploying a series of pumps to lift forty million gallons of waste per day to a height of fifty feet, to allow it to flow by gravity to the treatment works at Pigeon House.[27] The new facility was opened officially on 30 October 1986, fifteen months after it first came into operation, by Bertie Ahern TD, then the Lord Mayor of Dublin. The future Taoiseach marked the occasion with an idiosyncratic tribute to the contractors and workers who had performed an 'unglamorous but essential task'. Fergus Pyle, a former editor of the *Irish Times* who covered the event for the newspaper, gleefully recorded what the

---

24. Correspondence with Dr Brian Bond, 11 January 2010.

25. *Ibid.*

26. Interview with Martin Glynn, 16 June 2009.

27. *Irish Times*, 'Pat on the back for job done in Dublin's bowels', 31 October 1986.

politician had to say about the workers: 'One of them, he said repeating an old joke in the trade, had been overhead saying that the product might be excreta – the Lord Mayor used the shorter Anglo-Saxon term – "but it was our bread and butter".'[28] Lyal Collen, who also attended the ceremony, contented himself with a more conventional – and entirely accurate – commentary on the intricate engineering techniques involved in constructing the base of the pump house.[29] The pumping station at Ringsend ranked high among the most elaborate, innovative and difficult civil engineering projects completed by Collen Brothers. It was also one of the most prolonged contracts in the firm's history, surpassing even the long-running Portrane project at the turn of the century.

The completion of the new pumping station in some respects marked the end of an era for the firm, as the project was also the last of the large-scale civil engineering jobs undertaken by Collen Brothers. The firm concluded its involvement in the expansion of Dublin Port just before the pumping station became operational; with the completion of these important and lucrative projects, the firm's involvement in marine engineering also came to an end. This was not a matter of deliberate policy on the part of the directors. The opportunities for marine and other civil engineering contracts had largely dried up by the early 1980s, as the redevelopment programme for the port was brought to a conclusion, while the projects at Shannon and Aughinish were one-off developments.[30] Moreover, competition at tender stage became increasingly intense and the company failed to secure several marine engineering contracts in a highly competitive tendering process, including a new roll-on roll-off ferry terminal at Rosslare, which was instead won by Ascon.[31] The advent of recession, too, militated strongly against the initiation of further expensive projects in port or harbour development, either by the state or by major private investors. The marine work had accounted for most of the large-scale civil engineering projects undertaken by the firm over the previous generation, and the loss of such a central element of its business was a significant setback to the company.

The recession affected all aspects of the company's business, causing a significant decline in turnover for the first time in almost two decades and provoking an unusually severe retrenchment within the firm itself. The

28. *Ibid.*

29. *Ibid.*

30. Interview with Martin Glynn, 16 June 2009.

31. Interview with Pat Sides, 21 July 2009.

company was noted for its loyalty towards its staff, which was fully reciprocated by most of its employees.[32] Collen maintained a high level of direct employees up to this time and made considerable efforts to keep such workers on the books even during periodic dips in business activity. But on this occasion the severity of the recession triggered a far-reaching process of rationalization. The design office was the first to be affected, with over two-thirds of its staff being made redundant in the early 1980s, and its complement was reduced even further by the middle of the decade. Pat Sides recalled that: 'It was a tough time, it literally ended up with only Chris [Lyons] and myself in the design office.'[33] The impact of rationalization on the firm's capacity to deliver major civil engineering projects was particularly drastic. The departure of several experienced civil engineers in this period, through retirement or redundancy, meant that the company lost the expertise to undertake large-scale civil engineering contracts even if the directors wished to do so.[34] Civil engineering ceased to be a significant part of the firm's activity by the mid-1980s.

By 1983, the company as a whole had contracted dramatically with the workforce reduced by half.[35] A three-day week was introduced for most remaining staff, becoming an intermittent feature of working life within the company for much of the 1980s; pay cuts were also imposed at all levels.[36] It was a difficult period for the company, which lacked sufficient work to keep a substantial number of talented and experienced staff, and a traumatic process for those who were obliged to leave in the midst of a recession. It was almost as difficult for those who remained on short-time working hours.[37]

The radical changes within the firm during the 1980s were achieved with remarkably little internal conflict and virtually no strife between the proprietors and their employees. The Collen family business was a unionized company, which generally maintained good relations with union officials representing its employees. The family proprietors and senior managers maintained a strong attachment with their direct employees, many of whom spent their entire working lives with the company. The culture of the company under Lyal and

---

32. Interview with Paddy Wall, 28 April 2009; Interview with Des Lynch, 9 June 2009; Interview with Dr Brian Bond, 30 April 2009.

33. Interview with Pat Sides, 21 July 2009.

34. Interview with Dr Brian Bond, 30 April 2009; Interview with Paddy Wall, 28 July 2009.

35. Interview with Paddy Wall, 28 July 2009; Interview with Pat Sides, 21 July 2009.

36. Interview with Jimmy Small, 7 June 2009; Interview with Des Lynch, 9 June 2009.

37. Interview with Jimmy Small, 7 June 2009.

Standish incorporated a strong element of benign paternalism, in which the family directors expected (and usually received) the loyalty and commitment of their employees and in turn recognized obligations to look after their staff. Perhaps for this reason, the firm remained free of strikes or industrial disputes involving its direct employees, although it was sometimes embroiled in strikes affecting other contractors, as it had been at Aughinish in the early 1980s, or in wider disputes involving the construction industry as a whole. A long-serving member of the joinery shop commented wryly that: 'Although we were in trade unions, we weren't trade unionized'.[38] It was a tribute to the cohesion within the company and the strength of the bond between the proprietors and employees that the process of rationalization was accomplished without industrial relations strife. Indeed it was only much later, in the first decade of the twenty-first century, that the firm became involved in a significant industrial dispute. This was a major unofficial action by bricklayers, linked to the Building and Allied Trades' Union (BATU): the action was driven largely by resistance to the spread of subcontracting within the construction industry.[39] The dispute early in the new millennium remained very much an exception to the usual pattern of remarkable stability in Collen's industrial relations.

The firm attempted to manage the process to retain its most valued and knowledgeable employees, who were essential to the survival of the business. Des Lynch believed that the process of rationalization was very painful but ultimately necessary to preserve the business:

> They were really tough years and we all had to make sacrifices then. I remember I cut
> my own salary by one third to give an example to others; we had to do that just to get

---

38. Interview with Jimmy Small, 6 July 2009.

39. See pp.149–50 for the expansion of subcontracting; for the unofficial dispute in 2003 and 2006, see the following sources: *Irish Times*, 'Court order restrains pickets in subcontracting dispute', 14 November 2002; *Irish Times*, 'Injunction against pickets over sub-contractors continued', 29 November 2002; *Irish Times*, 'Construction row could deteriorate', 23 January 2003; *Irish Times*, '50 more builders laid off in Limerick', 25 January 2003; *Irish Times*, 'Judge hears of death threats against brick-layers', 28 January 2003; *Irish Times*, 'Builders due to resume work on Monday', 1 February 2003; *Irish Times*, 'Picketers of Dublin building site sent to jail', 11 February 2006; *Irish Times*, 'Union activity cause of work refusal, men claim', 11 February 2006; *Irish Times*, 'Jailed men due in court over picket', 13 February 2006; *Irish Times*, 'BATU official could not accept deal to end row', 18 February 2006; *Irish Times*, 'Union "main mover in intimidation"', 18 February 2006; *Irish Times*, 'Builders' union urged to help end protests', 23 February 2006; *Irish Times*, 'Carpenter agrees to stop picketing sites', 24 February 2006; *Irish Times*, 'Three picketers freed by High Court', 25 February 2006; *Irish Times*, 'Case against bricklayers is adjourned', 28 February 2006; *Irish Times*, 'Judge refuses order against building workers' union', 17 May 2006.

through. We were cut down to the minimum, we had very knowledgeable foremen that we tried to hold onto at all costs and then we selected the best of the tradesmen and they were to form the core … throughout the 80s and we managed to get through it.[40]

Pat Sides, who left the firm of his own accord in 1987 and returned two years later, gave a more blunt verdict: 'Standish moved fast, he knew what was coming, he was probably right in hindsight. But it didn't feel great at the time, it felt awful.'[41] Whatever its merits, however, the rationalization of the business was not restricted to its workers, but also extended to the highest managerial level and even the board of directors.

In the mid-1980s the senior proprietors moved decisively to transform the configuration of the family business. Standish and Lyal, the two main shareholders and still the dominant figures in the firm, acted to wind up the company, which they had created over thirty years before. An extraordinary general meeting of the shareholders on 20 December 1985 determined that Collen Brothers, Dublin would be placed in voluntary liquidation.[42] The proprietors used a newly constituted company to continue the business; Collen Construction Limited came into being in December 1984, with Standish and Lyal as its founding directors. The decision to wind up the old firm may have been influenced by personal tragedy within the family, namely the passing of Standish's wife: Claire Collen died in February 1985.[43] Claire was a minority shareholder within the firm and a formal valuation of its assets was undertaken for the first time following her death. While the initial steps to wind up Collen Brothers had been taken before Claire Collen's death, it is likely that her death influenced the timing of the process of liquidation.[44] The new company began to operate in the spring of 1985, gradually taking over the operations of the old firm so that no hiatus emerged in the business.[45] Significantly, the senior directors moved to appoint younger members of the family to the board of Collen Construction at its first meeting on 13 December 1984.[46]

---

40. Interview with Des Lynch, 9 June 2009.

41. Interview with Pat Sides, 21 July 2009.

42. Companies Registration Office, *Special Resolution of Collen Brothers (Dublin) Ltd. pursuant to Section 141 of the Companies Act, 1963*, 20 December 1985.

43. *Irish Times*, 'Personal', 25 February 1985.

44. Interview with Neil Collen, 28 April 2009.

45. Collen Construction Ltd, Minutes, Meeting of Directors, 13 December 1984; Minutes, 27 March 1985; Minutes, 17 May 1985.

46. Collen Construction Ltd., Minutes, Meeting of Directors, 13 December 1984.

David (left) and Neil Collen. Courtesy of Neil Collen.

Standish's son, David, was appointed as a director for the first time; so too was Neil Collen, Lyal's son, who had joined the company in 1978. The reconstitution of the company also led to a rationalisation of the board itself. The new company retained a smaller complement of senior managers, while the proportion of directors not bearing the Collen name was sharply reduced. John Griffin retired as managing director in 1984, while Brian Bond also departed from the company before the end of the year.[47] Dr Bond subsequently went into business independently and became a highly successful engineering consultant, as well as a part-time lecturer with the School of Engineering in Trinity College.[48] Four of the five directors on the new board were family members. Des Lynch, who became the first managing director of Collen Construction, was the sole director drawn from outside the family.

The voluntary liquidation of Collen Brothers was designed to allow the two senior proprietors to provide for their own retirement by realizing their assets in the firm. The special resolution, which was the final act of the old

47. Companies Registration Office, Certificate No. 13150/67, Form No.9, *Notice of change of directors or secretaries or in their particulars*, 1 December 1984.

48. Ronald Cox, *Civil Engineering at Trinity: A Record of Growth and Achievement* (Dublin, 2009), pp.108–9; Interview with Marcus Collie, 23 June 2009; Interview with Dr Brian Bond, 30 April 2009.

company, provided that the liquidator was 'authorized to divide and distribute among the members in specie any assets of the Company available for distribution.'[49] The board had adopted a policy of not declaring dividends for several years, with profits often being reinvested in the business and used to build up its assets: no dividend at all was paid to the shareholders between 1974 and 1982, despite the considerable success of the business for most of that time.[50] The two leading proprietors had deferred short-term gains over a long period to enhance the position of the firm. They had, however, taken the opportunity to secure personal investments in the firm's property portfolio, especially in the industrial estates at Tallaght and Bray.[51] Standish and Lyal took most of their shares out of the company in specie during the mid-1980s, by dividing up the property assets acquired in Tallaght and Bray. The senior directors acted to realize their long-term investment in the firm, while establishing a new company to ensure the survival of the business.

The proprietors set out to guarantee that Collen Construction had the necessary resources to operate successfully and did not have to meet outstanding liabilities from the previous incarnation of the firm. The new company acquired the builders' yard in East Wall, as well as all the plant and transport held by Collen Brothers, in December 1985.[52] Most of the remaining staff of the old firm were transferred automatically to the new company and the proprietors acted to meet the statutory obligations of Collen Brothers towards its employees. Standish and Lyal agreed to wind up the existing pension scheme and ring-fence the assets specifically for its beneficiaries among the firm's employees, with the actual payments to be made on retirement.[53] The proprietors also arranged that payments would be made from the old firm to the new covering accrued redundancy rights of all employees transferring from Collen Brothers.[54] The board resolved that staff transferring to the newly constituted company from Collen Brothers would be entitled

49. Companies Registration Office, *Special Resolution of Collen Brothers (Dublin) Ltd. pursuant to Section 141 of the Companies Act, 1963*, 20 December 1985.

50. Collen Minute book, Minutes, Annual General Meetings of Collen Brothers, 10 March 1977–15 December 1982; Interview with Paddy Wall, 28 April 2009.

51. Interview with Paddy Wall, 14 January 2010.

52. Collen Construction, Minutes, Meeting of Directors, 16 December 1985; Minutes, Meeting of Directors, 20 December 1985.

53. Collen Construction, Minutes, Meeting of Directors, 16 January 1985; Interview with Paddy Wall, 14 January 2010.

54. Collen Construction, Minutes, Meeting of Directors, 16 December 1985.

to an ex-gratia payment, in lieu of their service with Collen Brothers up to 31 December 1984, if they were made redundant by Collen Construction.[55] While the process of rationalization was largely implemented by this time, the provision made for redundancy rights underlined the bleak reality that further redundancies could not be ruled out and that job losses had become a normal part of working life in the 1980s. A significant trend in the reconstitution of the firm was the cautious approach taken by the senior directors in transferring funding to Collen Construction. The senior directors took care to ring-fence funding to meet specific requirements, such as pensions, redundancy or other statutory obligations. This approach reflected not only legal requirements and a genuine concern for the welfare of the company's employees, but also a distinct reluctance, particularly on Standish's part, to place the resources of the firm unconditionally at the disposal of a new generation of directors.

When Collen Brothers was placed in liquidation, a substantial proportion of the assets that had previously formed part of the firm's portfolio were taken out of the business at the instigation of the senior directors. At the same time the two men initiated a process of internal succession, not only delegating greater day-to-day responsibility to the younger family members but also transferring formal ownership of the new company. Standish and Lyal approved a new allocation of shares in Collen Construction, giving a majority stake to the younger family directors in December 1985. The senior directors retained only 20 per cent of the shares between them, while Neil and David secured a shareholding of 40 per cent respectively.[56] This transfer of ownership continued over the following decade: the younger directors took up further share allocations in April 1990, with the result that each man held 45 per cent of the company's shares, with Standish and Lyal each retaining a 5 per cent shareholding.[57] But this formal change in ownership did not entail an abrupt withdrawal by the dominant figures of the previous generation, nor did it mean that the younger members of the family were immediately given full control of the business. Neil Collen noted that 'I suppose there would always have been, an older generation fear that if the younger generation were given carte blanche and given everything on a plate, they would lose it fairly quickly.'[58] Yet the concerns of the senior proprietors should not be overstated, as they were essentially following

55. Collen Construction, Minutes, Meeting of Directors, 16 January 1985.

56. Collen Construction, Minutes, Meeting of Directors, 18 December 1985.

57. Collen Construction, Minutes, Meeting of Directors, 25 April 1990.

58. Interview with Neil Collen, 28 April 2009.

the lead established by their own predecessors. Internal succession had always been a gradual process within the Collen family business.

The board of Collen Construction operated as a collective institution in which all four family directors had a voice, but the senior directors retained considerable influence in the running of the firm. Standish remained chairman of the board for over eleven years and continued to exercise oversight over the management of the business for much of that time. More crucially, the involvement of both senior figures was indispensable to the successful operation of Collen Construction, as they held a substantial segment of the assets previously vested in the firm. The backing of the senior directors was vital due to a long-term requirement within the construction industry for building contractors to issue bonds to guarantee contracts: the contractor was frequently required to take out an insurance bond amounting to 25 per cent of the contract value, which inevitably meant the pledging of assets as collateral. The issuing of bonds was mandatory for all public sector contracts and for private work wherever it was sought by the client, although some long-term clients of the firm were often willing to waive this requirement.[59] Standish and Lyal provided the necessary security to allow the company to enter into contract guarantee bonds, ensuring that both men retained a substantial input on key decisions affecting the business.[60] The establishment of Collen Construction began a gradual transition to a new generation, but did not signal that the senior directors were immediately relinquishing their central position in the company. The changing of the guard was not completed until 1996 and was then accompanied by a further change in the ownership of the company.

The decisions taken by the family directors in the mid-1980s brought the most radical transformation of the company since the division of the old firm over a generation earlier. The traditional brand name of Collen Brothers was consigned to history without any obvious soul-searching on the part of the proprietors. Collen Construction emerged as the new commercial vehicle to continue the family business. The reconstitution of the firm formed part of the family's response to the economic pressures generated by the recession, producing a smaller company with a reduced managerial cohort. Yet it was primarily the closely related issues of retirement and succession that dictated the formation of the new company. The senior proprietors took an unsentimental decision to wind down Collen Brothers, Dublin, because they aimed

---

59. Interview with Paddy Wall, 14 January 2010.

60. Interview with Neil Collen, 28 April 2009.

to provide for their retirement on a phased basis and to safeguard their assets. The management of internal succession, a consistent preoccupation of senior directors since the creation of the original firm over a century before, was also a key motivation in the decisions that led to the emergence of Collen Construction. The initiatives taken by the senior directors reflected their concern to begin a process of transition to the next generation, while also indicating a degree of caution about handing over full authority to younger family members immediately.

The evolution of the firm was also shaped by wider developments within the construction industry as a whole, notably the gradual replacement of direct employees by subcontractors. The company relied to a considerable extent on direct employees up to the 1970s, in line with traditional work practices. While the firm employed subcontractors to undertake plastering, all the other work on its sites, including carpentry and bricklaying, was carried out by direct employees.[61] Moreover, the fitters' shop and joinery shop in East Wall Road were staffed entirely by the company's own employees. The traditional pattern began to change in the 1970s, as the lower costs of subcontracting became more attractive to employers. Yet senior company insiders estimated that between 50 per cent and 60 per cent of the firm's turnover in the late 1970s was still carried out by direct employees.[62] The recession in the early 1980s accelerated the transition from directly employed labour to subcontracting. The wave of redundancies removed a significant proportion of direct employees, who were never replaced. Instead the company turned to subcontractors for essential skilled work on building sites when economic conditions began to improve towards the end of the decade. The vast majority of the workers employed by Collen on building sites were subcontractors by the end of the twentieth century; only the foremen and a small number of general operatives were still employed directly by the company.[63] Neil Collen noted the sweeping nature of the transition from direct employment to subcontracting from a contemporary vantage point in 2009:

> So [in the 1970s] if there were 40 people on a site, we would have employed, say, 20 of those, whereas today if there were 40 people, because of the way the building industry has evolved over the last 25 years, we might only have three or four directly employed.[64]

---

61. Interview with John Ruane, 16 July 2009.

62. Interview with Neil Collen, 28 April 2009.

63. Interview with John Ruane, 16 July 2009; Interview with Marcus Collie, 23 June 2009.

64. Interview with Neil Collen, 28 April 2009.

This trend did not have a uniform impact throughout the company, as it affected mainly the skilled trades and the employment of labour on construction sites. The management and staff in the company's offices remained direct employees of Collen Construction. Yet the transition to widespread employment of subcontractors marked a far-reaching transformation in the nature of the construction industry, which inevitably exerted a profound influence on the operation of the company.

A less dramatic but equally important change was the gradual introduction of information technology. Collen Brothers did not employ computers at all during the most successful period of its operations between the mid-1960s and late 1970s. While correspondence was tapped out on manual typewriters, most of the records generated by the firm, including the accounts and pay roll for wages, were produced by hand. Paddy Wall recalled his own experience in the wages office during the 1960s: 'absolutely everything was done by hand. The accounts were worked out tediously in long hand and many was the long night I spent with Bill McCullough doing wages or sometimes helping him to prepare graphs or statistics for Standish.'[65] Similarly, the documents for contract tenders, especially the bills of quantities that set out the detailed prices for a contract, were produced by hand, usually under the direction of Roger Beckett, the company's chief quantity surveyor. Beckett showed an impressive ability to price jobs meticulously without the assistance of modern technology, as Paddy Wall noted:

> The bills of quantities were always completed initially in pencil because changes might be made in the details. But at the eleventh hour somebody had to ink in all the details in all those columns. Roger was quite happy to work out prices for jobs and he was very good at it and very quick.[66]

The status quo persisted until the late 1970s, when the accounting section took the lead in adopting modern technology. Marcus Collie, a business graduate of Trinity College, who had married Standish's daughter, Diana, played a key role in introducing the firm to the benefits of information technology. He joined Collen in 1978, acting initially as the accountant on the joint venture project with Christiani & Nielsen at Aughinish. He took over as the company's management accountant when William McCullough died in 1979.[67] Marcus Collie set out to modernize the firm's financial management and accounting system: 'All the book

---

65. Interview with Paddy Wall, 28 July 2009.

66. *Ibid.*

67. Interview with Marcus Collie, 23 June 2009.

keeping was done by hand, which even to me was absolutely ridiculous and so the first two things I did when I got here, I was here about three months and I got a computer and a coffee machine. The jury is still out as to which was the more valuable asset to the company!'[68]

The company's first computer was a vast ICL, which evoked awe and bemusement in almost equal measure among the more experienced staff. Several senior employees remained attached to more traditional methods. Roger Beckett never embraced information technology and continued to work entirely by hand.[69] But the younger office staff welcomed the introduction of computers, which did much to simplify their workload. Rita McMillan, who worked in the accounts department during the 1980s, welcomed the greater efficiency of the new system, even if the new computer had its fair share of quirks: 'It was a huge thing, it had its own room. And it saved an awful lot of time because you weren't handwriting everything. But it was finicky … if it was too cold it wouldn't start, if it was too hot it used to skip things in print-outs, so it was kind of temperamental.'[70]

The advent of information technology transformed the working life of the company over the following decade and a half. Yet the transition to the information age did not happen overnight. When Louise Coffey first joined the company as a receptionist in 1989, she undertook a typing exercise as part of the interview process, 'typing on this really old typewriter'.[71] The adoption of new working methods, which made full use of computing, occurred gradually, due in part to the persistence of traditional preferences among a minority of staff but also as a consequence of financial constraints for most of this period.

The use of information technology became a normal feature of working life throughout the company by the early 1990s. The design office switched over to computers at the beginning of the new decade, dramatically reducing the time required to design a project before it reached the construction site, and allowing the design team to complete their work in about a third of the previous time.[72] Frank O'Sullivan, a senior architect whose experience spanned over forty years within the company, singled out the adoption of information technology as the most vital change during his tenure:

---

68. *Ibid*.

69. Interview with Louise Coffey, 23 June 2009.

70. Interview with Rita McMillan, 13 July 2009.

71. Interview with Louise Coffey, 23 June 2009.

72. Interview with Frank O'Sullivan, 1 July 2009.

The introduction of computerisation changed everything and we would have been one of the first to be involved in computer drafting, computer-aided design and drafting. You came from a situation where you might have six months of manual drafting to do the job, to design it … When computerization came, you know everything was fast, that six months would have shrunk down to maybe two months and it would certainly coincide with the planning period [of three months]. So while it was in for planning you had the capacity to design everything and get it ready for site at the same time.[73]

The Collen design team was one of the first in Ireland to employ computer software for design purposes. The company's design staff took full advantage of the latest innovations in computing and the emergence of advanced software from the early 1990s. It was a far cry from the handwritten tender documents and elderly typewriters, which were ubiquitous within the workplace only a short time before. The introduction of information technology entailed a gradual modernization of the company's work practices, which was essential in adapting the firm to a rapidly changing business environment and enabling Collen to compete effectively within the modern construction industry.

Odlums' Flour mill, Kennedy Quay, Cork city. Courtesy of David Collen.

73. Interview with Frank O'Sullivan, 1 July 2009.

While senior managers within the company were certainly conscious of long-term trends such as the prevalence of subcontracting and the process of modernization, they were largely preoccupied with the short-term demands of sustaining the business, by no means a simple undertaking during the 1980s. The loss or decline of major sources of business greatly influenced the activity of Collen Construction and made the company more dependent on straight-forward building projects, which had formed the largest part of its turnover up to the 1960s. The company's ability to nurture long-term client relation-ships once more provided a crucial element of its business, with Odlums in particular offering repeat work. Collen built a new flour mill in Cork city for Odlums during the mid-1980s, with the work extending through four phases and over a couple of years. David Collen, who supervised part of the project, was relieved at the relatively benign economic conditions that appeared to prevail in the construction industry in Cork:

> I would have been down to Cork, to Odlums … Then [the work in] Cork took on a very interesting life, things were very bad up here [Dublin] in the early '80s, fellas getting laid off left, right and centre. We couldn't believe it down in Cork, we were in a cocoon down there. I went down to snag phase 1 or 2 of the job, as I was down there, the lads were getting phase 3 and 4.[74]

But Collen failed in tendering for other projects in Cork around the same time: the firm's success in this instance was due less to local economic condi-tions than to long-term personal and professional connections with Odlums. The company secured the contract for the mill in part due to its successful delivery of similar projects stretching back to the 1950s, but also because Lyal maintained close connections with the owners of Odlums, another family business founded in the mid-nineteenth century.

The company also began to acquire a reasonable share of public building contracts during this period. Collen built a new recreation hall in Finglas for the City of Dublin Vocational Education Committee during the mid-1980s – the project with the VEC was worth just under £1 million in 1985.[75] The company also completed a range of building projects for the Office of Public Works (OPW). Collen fitted out new units for Ballymun town centre in 1985 and undertook a more substantial job in the following year with the extension of Agriculture House in Kildare St, which included the provi-sion of a computer room, plant rooms and accommodation for welfare

---

74. Interview with David Collen, 10 June 2009.

75. Collen Construction, Minutes, Meeting of Directors, 3 July 1985.

officers.[76] As a medium-sized firm, Collen rarely won large-scale state contracts, which normally went to the most prominent building contractors, such as Sisk or Crampton. But the company secured a valuable share of the business offered by local authorities and the OPW, which increased substantially as economic conditions improved over the following decade.

The recession did not prevent a new departure by the company, which engaged in residential development for the first time. Collen had built houses and to a more limited extent, housing estates, in the first half of the twentieth century, but its involvement in the housing market ended by the early 1950s. Despite the company's established track record in the development of industrial property, Collen had no involvement in the development of houses and apartments until the 1980s. This was a deliberate decision on the part of the senior proprietors, not least because Standish was profoundly sceptical of the value of house building. Pat Sides recalled, 'Standish always said you never make money building houses.'[77] The company's first venture in residential development came in 1984, on its own land at Kingston Lodge on Clonliffe Road, Drumcondra. Kingston Lodge was an old Victorian residence, which was owned by the Collen family and was Harky Collen's home for much of the twentieth century.[78] The house continued to be a family residence until it burned down in a fire. The company developed a new apartment complex on the same site. The first phase of the development consisted of twenty-four one-bedroom apartments, which were built in two blocks.[79] The first two blocks were ready for occupation in the spring of 1984, but the company had sold only nine of the new apartments by the following July.[80] The firm secured planning permission for a second phase of apartments, but did not go ahead with the building project and later sold off the remaining sites at Kingston because demand for the original apartments was so limited. The development at Kingston was a modest initial venture by the company into residential property and apartment building. It had no immediate sequel, as the depressed property market of the period did not provide fertile ground for development. The fact that Collen initiated only a single housing project speaks volumes

76. Collen Construction, Minutes, Meeting of Directors, 25 September 1985; Collen Construction, Minutes, Meeting of Directors, 17 September 1986.

77. Interview with Pat Sides, 21 July 2009.

78. *Irish Times*, 'Compact development: Nine sold at Kingston', 13 July 1984; *Memorandum and Articles of Association of Collen Brothers (Dublin) Ltd.*, p.11.

79. *Irish Times*, 'First show flat to open at Kingston', 3 February 1984.

80. *Irish Times*, 'Compact development: Nine sold at Kingston', 13 July 1984.

about the firm's attitude to residential development: between 1980 and 1995 the company's management remained cautious, with good reason, about the potential of housing development at least until the mid-1990s.

Following a steep decline in turnover and profits during the mid-1980s, the company's fortunes fluctuated in accordance with wider economic trends throughout the following decade. The construction industry enjoyed an upturn in activity by 1989, with the initiation of new housing developments in Dublin and a revival in the demand for industrial units.[81] Collen Construction had no involvement in providing such housing estates, but took the opportunity to move back into industrial development on its old stamping ground in Tallaght. The company developed four new industrial units at Broomhill Terrace in 1989, mainly devoted to warehouse and production space, but also incorporating office accommodation.[82] Three of the four units were leased almost as soon as they came on the market in January 1990.[83] Collen also took charge of the construction of new industrial units for Friends Provident on the same estate. The insurance company aimed to create property investments in Tallaght for their pension fund – Collen sold the land to Friends Provident and secured the contract to design and build the new accommodation, which was marketed as the Broomhill Business Park. The firm provided eleven large industrial units, offering space for offices and warehouses, which ranged in size from 5000 to 31,500 square feet.[84] The revival in the company's business was reflected in its accounts. Collen Construction returned a profit of £10,118 before taxation in 1989–90, compared to a loss of £4411 in 1988–9.[85] The company enjoyed a dramatic advance in the following year, benefiting especially from industrial development activity in Tallaght and Bray. Collen recorded a pre-tax profit of £112,570 in 1990–1, over a tenfold increase within a single year.[86] But this

---

81. *Irish Times*, 'Construction industry gearing up for period of high activity', 30 December 1989.

82. *Irish Times*, 'To Let, Broomhill Terrace, Tallaght industrial estate, Dublin 24', 10 January 1990.

83. *Irish Times*, 'Rents for industrial units rise by 50%', 28 February 1990.

84. *Irish Times*, 'Tallaght business park to be ready by next year', 20 September 1989; *Irish Times*, 'Rents for industrial units rise by 50%', 28 February 1990.

85. Companies Registration Office, Collen Construction Ltd, *Profit and Loss Account for the year ended 31 March 1990*, p.6.

86. Collen Construction Ltd, Accounts, *Profit and Loss Account for the year ended 31 March 1991*, p.6.

revival, although impressive in itself, proved a false dawn rather than an immediate portent of better things to come.

The pace of recovery within the wider economy was slow and uneven. The modest revival of the Irish economy at the end of the 1980s was stalled by a further recession early in the following decade, with unemployment in the Irish state reaching a historic peak at 20 per cent of the labour force in 1992. Collen Construction suffered heavily from the renewed downturn, registering a significant loss amounting to almost £200,000 in 1992–3.[87] The company continued to struggle during the following year, undergoing a further decline in turnover, but returning a smaller loss due to reductions in costs.[88] The reduction in overheads was achieved by a temporary return to the three-day week in 1994, combined with voluntary pay cuts taken by senior staff.[89] The gradual improvement in economic conditions by the mid-1990s, which ultimately foreshadowed the emergence of an economic boom, helped to restore the construction activity on which the company depended. Collen returned to profitability by March 1995, although the respectable profit achieved by the company in 1994–95 was still considerably lower than it had been at the beginning of the decade.[90] The company experienced distinctly mixed fortunes in the context of volatile economic conditions, with a reasonable level of activity at the beginning of the 1990s closely followed by a further decline in business and revenues between 1992 and 1994. The moderate recovery achieved by the company in the mid-1990s would soon provide the platform for more dramatic advances in an era of unprecedented expansion.

The transition within the business from the senior proprietors to the younger family members was an ongoing theme for over a decade, from the establishment of Collen Construction until the mid-1990s. The two senior directors gradually withdrew from day-to-day involvement in managing the business during the late 1980s, although they retained a pivotal position on the board and continued to influence key decisions. While Standish and Lyal still attended board meetings regularly, Des Lynch took responsibility

---

87. Collen Construction Ltd, Accounts, *Profit and Loss Account for the year ended 31 March 1993*, p.6.

88. Companies Registration Office, Collen Construction Ltd, Accounts, *Profit and Loss Account for the year ended 31 March 1994*, p.6.

89. Interview with Rita McMillan, 13 July 2009; Interview with Louise Coffey, 23 June 2009; Collen Construction, Minutes, Meeting of Directors, 17 April 1996.

90. Collen Construction Ltd, Accounts, *Profit and Loss Account for the year ended 31 March 1995*, p.6.

for the everyday management of the firm, assisted by the two younger directors. When Des Lynch retired in February 1993, Neil and David Collen took over the management of the company.[91] The collaboration between Lyal and Standish in the previous generation appeared to provide a viable precedent for the two younger directors. Yet the latest dual managerial arrangement lasted for barely three years. Both the external environment and internal dynamics within the company were very different from thirty years before. The uncertain economic context in which the company operated since the early 1980s placed a premium on the ability to reach key decisions rapidly and decisively; it was not yet apparent that Ireland was about to enter a sustained period of economic prosperity. But the internal managerial arrangement, allied to the continuing requirement for contract guarantee bonds, tended to militate against effective decision-making. In particular the need to pledge personal assets, held by the two younger men or their older relatives, as collateral for contracts gave each director a veto on the activity of the other.[92] It became apparent that the joint managerial arrangement would not be sustainable in the long term. A consensus gradually emerged within the Collen family that one of the younger directors would take full control of the company and that one branch of the family would buy out the other's interest in the business. Neil and David were determined to keep the company together as a single unit. Both men shared a common view of the firm as an institution whose importance went beyond the interests of the family, but was equally central to the welfare of its employees, many of whom had served the company for over thirty years.[93]

Following negotiations among the directors, Neil Collen reached an agreement with his uncle and cousin to buy their shares in the spring of 1996. A legally binding resolution passed by all four directors on 17 April 1996 recorded the agreement between them and determined the future ownership of the company. Neil and Lyal purchased the full shareholding of their relatives for £300,000 – both Standish and David resigned as directors of the company on the same day.[94] Lyal was appointed as chairman of the board, while Neil took over as managing director of Collen Construction; it was generally acknowledged

---

91. Collen Construction, Minutes, Meeting of Directors, 3 March 1993.

92. Interview with Neil Collen, 28 April 2009.

93. Interview with Paddy Wall, 28 April 2009; Interview with Neil Collen, 28 April 2009; Interview with David Collen, 16 June 2009.

94. Collen Construction, *Resolution of Directors passed pursuant to the Companies Act 1963*, 17 April 1996, p.1.

that the control and management of the firm had passed to the younger man.[95] The agreement resolved the complex issue of succession within the firm, which had preoccupied the directors and been a cause of concern to many employees since the dominant figures of the previous generation signalled their intention to retire over a decade previously. The gradual process of transition that began with the winding up of Collen Brothers in 1984 concluded over eleven years later when Neil Collen took sole control of the business.

The generational transition that unfolded during this period was undoubtedly prolonged and occasionally convoluted. Yet the senior proprietors arguably made the right decision, even if they did so inadvertently. The prolonged nature of the transition proved beneficial because it gave time and space for consensus to emerge among the family directors on the way forward for the company. The process of succession was ultimately accomplished in an amicable fashion and on the basis of a mutually beneficial agreement. The ability of the Collen family to resolve the issue of internal succession in an effective fashion was a key element in the survival of the business. The agreement between the directors in the mid-1990s did much to illuminate not only the *modus operandi* of the family proprietors but also the character of Collen as a business institution. Whatever differences arose between the proprietors, they were united in their determination to maintain the company as a family business and to keep it intact as a single organization. The priorities of the directors reflected their sense that the company was not merely a profit-making instrument but a social community, with reciprocal obligations between directors and employees.

Such a sense of mutual obligation within the firm did not override commercial realities. The senior proprietors presided over a painful process of rationalization during the early to mid-1980s, which brought a substantial enforced exodus from the company and imposed considerable sacrifices on those staff who remained in place. Yet the sometimes controversial decisions of the family directors did not change the genuine sense of common purpose within the firm nor did they affect the loyalty that the company inspired from its employees. Collen had evolved a culture of community and mutual support between directors, managers and employees that survived the severe shocks of the 1980s – this ethos reflected Collen's essential character as a family-run enterprise and did much to explain the preservation of the business during the most challenging economic circumstances since the 1930s.

95. Collen Construction, Minutes, Meeting of Directors, 18 April 1996.

SIX

Transforming the Business

The two independent companies, which emerged in 1949, shared a common heritage and at least one defining characteristic – the enduring character of the business as a family-run institution. Yet in almost every other way the firm in Portadown evolved in a very different fashion to the newly constituted enterprise in the Republic. Following the agreed separation in 1949, the family shareholders in Portadown kept control of the original firm, which retained its existing corporate status and continued to trade as Collen Brothers Ltd. The company continued, initially at least, the range of commercial activities promoted by previous generations of the family in Portadown. The northern branch of the firm showed apparent parallels with its southern counterpart in the period immediately following the division of the business, but such common features largely disappeared as the two companies went their own way during the post-war era. Collen Brothers in Portadown had to contend with the Troubles, which not merely presented commercial difficulties but also posed serious personal risk to directors and employees. Yet the grim background of the Troubles did not determine the distinctive path taken by the company, which was dictated by economic realities, the traditions of the firm itself and the commercial decisions of its leading members.

Jack Collen (in the foreground), with Joe Collen, overseeing the work on site.
Courtesy of Peter Collen.

Collen Brothers rarely had difficulty in living up to its name and the original company in Portadown proved no exception. The firm, not for the first time in its eighty-year history, came under the direction of two brothers, in a striking parallel with the practice adopted south of the border. Tony and Joe Collen, who had already taken a key role in managing the business before the Second World War, were the leading figures within the company for over a quarter of a century after 1949. The two men were the sole directors and shareholders until the 1970s, with the share capital of the company being divided equally between them.[1] They were not simply equal proprietors but operated jointly in managing the business, and neither assumed the dominant position held by their grand-uncle John within the original firm or the key managerial role established by Standish during the same period in Dublin. Instead the family directors in Portadown established a functional division of labour, with Joe taking responsibility for building projects along with the hardware and builders' providers company, while Tony took charge of quarrying and civil engineering, including road construction.[2] Several senior employees played a

1. Interview with Niall Collen, 5 February 2010.

2. Interview with Niall Collen, 14 May 2009.

Tony Collen (on right). Courtesy of Eleanor Gough (née Collen).

vital part in running the firm, notably Willie Eakin, who worked closely with Joe Collen in pricing jobs and company administration, Victor Whitcroft, a Trinity graduate who was the senior engineer on the civil engineering side of the business and Dick Gilpin, the company buyer.[3] Another key figure was Victor Gilland, who originally joined the firm in 1938 and served as manager of the quarry for much of the post-war period.[4] The company also relied heavily on a group of capable and experienced foremen, who supervised the work on the construction sites. Joe Collen later recalled that 'we were blessed with some very good men who worked for us then.'[5] The directors in Portadown maintained a distinctive managerial approach, which was characteristic of a traditional family business, delegating responsibility to a small number of key employees, but not elevating senior managers to the board itself. The two directors established a joint managerial model, which served the firm well for a generation.

The extension of the post-war welfare state to Northern Ireland and the willingness of the British government to finance increased spending on the province's social infrastructure contributed to a benign economic environment for the construction industry, facilitating the company in securing extensive building work between the late 1940s and the early 1970s. The

3. Correspondence with Niall Collen, 16 March 2009.

4. Collen Papers Portadown, Minute book, Collen Brothers (Quarries) Limited, Minutes, Annual General Meeting, 17 August 1984.

5. Interview with Joe Collen, 14 May 2009.

Northern Ireland government and local authorities undertook substantial capital expenditure, which was largely subsidized by Whitehall during the two decades following the war – such investment stimulated the expansion of public housing, as the Stormont administration established the Northern Ireland Housing Trust to provide houses financed by the state and introduced generous state subsidies for house-building.[6] Collen, like many other builders in Northern Ireland, profited from the determination of the Unionist establishment to secure parity of services for the province with the rest of the United Kingdom. The company also benefited, indirectly but significantly, from the considerable largesse of the British taxpayer. The firm's central role in building the Woodside estate in 1946 marked the beginning of a prolonged and highly successful engagement by Collen with public housing developments in Northern Ireland.[7] The first major project undertaken by the company following the division of the business was a new housing scheme at Killycomain, close to the site of the house once owned by John Collen, the founder of the firm. Collen Brothers built two hundred houses at Killycomain for Portadown urban council in the early 1950s.[8] Although Collen had built public housing before the Second World War, the post-war developments undertaken by the firm were much larger in scale and consisted of higher-quality housing stock. While terraced housing was the norm for public schemes during the 1930s, the estates at Woodside and Killycomain were built in blocks of semi-detached houses.

The company benefited too from public investment in school building, especially at secondary level, which had largely been neglected by the state before the Second World War.[9] Collen was engaged in educational projects in Portadown itself, building a secondary school at Clounagh in the 1950s and a primary school at Bocombra during the late 1960s.[10] The job at Clounagh was memorable for those involved as it saw the use of a tower crane by the firm for the first time. The company also took charge of the construction of a new secondary school at Toberhewney Lane in Lurgan during the 1950s, while building technical schools in both Lurgan and Armagh during

6. Lyons, *Ireland since the Famine*, pp.741–4; Ferriter, *Transformation of Ireland*, p.450.

7. See Chapter 3, p.66.

8. Collen Papers Portadown, Speech by Joe Collen to the Rotary Society, *My Job*, March 1971, p.3.

9. Lyons, *Ireland since the Famine*, p.742.

10. Interview with Niall Collen, 5 February 2010.

Opening ceremony for Technical School at Lurgan, 1961, undertaken by
Prince Philip, the Duke of Edinburgh. Courtesy of Peter Collen.

the following decade. The education department of Armagh County Council
was the client for most of the educational projects undertaken by Collen and
the firm's involvement in school building was concentrated almost entirely in
Co. Armagh.

Road surfacing and construction emerged as a central feature of the company's business in the 1950s, maintaining an involvement in quarrying which was
a traditional element of the firm's activity since the early 1900s. Following the
closure of the firm's quarry at Killycomain shortly after the Second World
War, Collen acquired the use of a new quarry on the Duke of Manchester's
land at Tandragee in 1948, paying both rent and royalties on output from the
quarry to the Duke's agent.[11] The ready supply of raw materials from the quarry
contributed to the firm's success in being awarded numerous road reconstruction schemes. The company began to manufacture bitumen macadam, having
acquired one of the first mixing plants in Northern Ireland during the mid-
1950s. The availability of quarry products allowed the firm to become a semi-
permanent contractor for road surfacing with the local authorities.

11. *Ibid.*

The first bitumen macadam mixing plant, acquired by Collen Brothers, 1955.
Courtesy of Peter Collen.

Collen Brothers was constantly engaged on contracts with Armagh County Council for resurfacing work from the 1950s. The firm's employees were a regular presence on the roads for much of the year, laying bitumen macadam. Niall Collen recalled that 'the paving squad would have been out continuously throughout the better half of the year.'[12] Moreover, the quarry was a crucial asset in enabling the company to undertake large-scale road construction projects. One of the first major reconstruction projects completed by Collen was the upgrading of the road between Armagh and Portadown in the 1950s. The following decade, however, marked the high point of the firm's involvement in road construction. Collen upgraded the Markethill-Armagh route in 1967, reconstructing about eight miles of the existing road: it was the largest road construction job completed by the company up to that point, which was valued at £1 million.[13] While Armagh County Council was undoubtedly the major client for road construction projects, the company also secured similar contracts from other local authorities. Collen upgraded two sections of the A1, the main Dublin-Belfast road, one at Banbridge and the other near Newry, during the 1960s.[14] The road construction and maintenance projects provided an invaluable element of the firm's turnover in the early post-war period.

12. *Ibid.*

13. Speech by Joe Collen, *My Job*, March 1971, p.3.

14. Interview with Niall Collen, 5 February 2010.

Road works: Collen Brothers' site at Drumlyn crossroads, 1963.
Courtesy of Peter Collen.

Knock bridge over the River Bann being constructed by Collen Brothers, with
Joe Collen's house in the background. Courtesy of Peter Collen.

The company's extensive involvement in large-scale civil engineering projects accelerated a gradual mechanization of its plant. The firm still used horse-drawn transport to a limited extent in the 1940s, but the horse and cart soon became redundant, due to an increased volume of work, technological advances in machinery and the greater demands posed by large-scale road construction or building contracts. The firm first acquired heavy dozers from the US army in Britain, which had a surplus of heavy machinery following the end of the war.[15] Collen bought three new push scrapers to move large quantities of earth on the Markethill-Armagh road: the new Terex TS–14 machines were essential to the effective completion of the project, along with new International Harvester TD–25 dozers that were purchased from the USA at the same time.[16] The mechanization of plant in the post-war era reflected an international trend throughout the construction industry. Joe Collen identified the eclipse of the horse and cart and the trend towards the adoption of more advanced machinery as key changes during his long career with the firm.[17] Mechanization was completed by the 1960s, although the modernization of plant remained an ongoing challenge for the firm.

The establishment of a new city in north Armagh, named Craigavon after the first Prime Minister of Northern Ireland, created wide-ranging opportunities for Collen Brothers. An official report composed by Sir Robert Matthew, the *Belfast Regional Survey and Plan*, published in February 1963, proposed the creation of Northern Ireland's first new city – the plan envisaged that Portadown and Lurgan would be linked within the new urban centre.[18] The initial scheme proved too ambitious and was not fully achieved, but a new urban centre was developed between the existing towns, under the auspices of the Craigavon Development Commission. The new city gradually took shape between 1965 and 1973: as a major local building contractor, the company was ideally placed to benefit from the largest urban development in north Armagh since the industrial revolution. Collen took responsibility in 1969 for a series of large-scale housing projects in Craigavon, building approximately 500 houses over a period of several years.[19] The company built public housing estates for the Craigavon Commission in the townlands of Drumgor,

---

15. *Ibid.*

16. *Ibid.*; Speech by Joe Collen, *My Job*, March 1971, p.3.

17. Speech by Joe Collen, *My Job*, March 1971, p.3.

18. Lyons, *Ireland since the Famine*, p.747.

19. Speech by Joe Collen, *My Job*, March 1971, p.3.

Enniskeen and Moyraverty within the emerging urban centre.[20] The successive phases of development in the new city marked the most extensive involvement by the firm in house building, on either side of the border, since the partition of the island. Moreover, Collen was not slow to exploit the opportunities offered by the far-reaching scheme of urban development for its civil engineering business. The Craigavon Commission awarded a contract to the company for the provision of a major roads scheme in the centre of the new city. The firm constructed the entire road and bridge network at the heart of Craigavon during the early 1970s.[21] The Commission proved one of the most valuable clients acquired by Collen during the post-war era, although the development body's life span was relatively short and it was replaced by a new borough council in 1973. The company found in the new city of Craigavon a rare opportunity to serve the two most significant strands of its business simultaneously, undertaking substantial building and road construction operations on the same urban site.

The Collen family enterprise also incorporated a builders' providers and hardware business, already a fixture of the original firm before 1949. A subsidiary company, Collen Brothers (Builders Providers) Ltd, conducted this strand of the business during the post-war era; it was based in Hanover St, beside the offices and workshop of the main company. An advertisement in the *Portadown Times* in May 1962 announced that the operation would supply 'all Materials to build your New Home or Renovate the Old One … in fact everything for the Builder.'[22] The builders' providers company supplied not only materials such as bricks and concrete blocks but also a wide range of finished products for house building, including bathroom units, doors and windows.[23] The company also sold an impressive variety of hardware, including ranges to heat houses before the advent of central heating, farm gates and water pumps.[24] The builders' providers operation remained a significant part of the family firm for two decades after the Second World War. But the subsidiary company ceased to trade in the 1960s, and the proprietors decided to end their involvement in the business. While the reasons for their decision are not recorded, the profits of the business had recently declined and it is likely that the proprietors decided to cut their losses. Collen Brothers sold the business

---

20. Interview with Niall Collen, 5 February 2010; Interview with Joe Collen, 14 May 2009.

21. Interview with Niall Collen, 5 February 2010.

22. *Portadown Times*, May 1962.

23. *Ibid.*

24. Interview with Niall Collen, 5 February 2010.

to a local supermarket, Henry Emerson – the building that accommodated the hardware store was later burned down in an accident.[25] The closure of the builders' providers company underlined the willingness of the proprietors to abandon long-standing elements of the business where it made commercial sense to do so. The pragmatic calculations of the directors would lead to a wider restructuring of the firm itself in the following decade.

The passionate commitment given by the partners to the family business was vividly illustrated in a speech by Joe Collen to the local Rotary Society in March 1971. Having given a brief sketch of the history of Collen Brothers since 1867, the long-serving director placed his own and his brother's career firmly in a broader context:

> We are the fourth generation in the construction business and I hope there will be a fifth. I am a builder, I was born into it and grew up in it and am glad that circumstances never forced me to enter any other occupation. For me it has been completely satisfying, one can always go back to various sites and see the fruits of one's labour.[26]

Joe Collen took pride in the tangible achievements of the family business and identified his own generation as the heirs of a building tradition, which traced its origins back to the stonecutters of Tandragee in the early nineteenth century. He expressed confidence that his son Niall would follow him in the construction business: 'he has been brought up in the building atmosphere and it would be almost impossible for him not to become involved'.[27] Niall graduated from the School of Civil Engineering in Trinity College in 1971 and worked for a year with Taylor Woodrow in London. It was an outbreak of industrial action, a common occurrence in British industry throughout the 1970s, which triggered his return to Portadown: 'There was a strike in the construction industry in England at that stage. I remember them [Joe and Tony] coming on the phone and saying there's no sense in you hanging around doing nothing over in London, you would be better off back here working in the business.'[28]

Niall began his career in Collen Brothers as a junior engineer, setting out the plans for building sites. His apprenticeship within the business proved relatively short, however, as the company experienced a far-reaching transformation within a few years of his return to Portadown.

---

25. Interview with Niall Collen, 5 February 2010.

26. Speech by Joe Collen, *My Job*, March 1971, p.3. In fact, Joe and Tony were the fifth generation to be engaged in the building trade since John Collen Snr in the early nineteenth century.

27. *Ibid.*

28. Interview with Niall Collen, 14 May 2009.

Niall Collen. Courtesy of Peter Collen.

Despite their undoubted pride in the company's achievements, the proprietors were rarely sentimental about the nature of their work, nor were they particularly attached to any corporate structure. Their pragmatism came to the fore in the increasingly volatile economic conditions of the 1970s. The rate of inflation in the UK increased rapidly for much of the decade, reaching 20 per cent by 1975. The steady escalation in prices caused severe problems for Collen, especially in its building operations. The company was committed to 'fixed-price' contracts in the building business which could run for up to three years. The direct costs for building projects, especially wages, were constantly increasing, while the company remained committed to a specific price for the completion of the job.[29] The inflationary spiral meant that it was extremely difficult to obtain a reasonable return from building projects. Niall Collen recalled that it was virtually impossible to price jobs accurately or predict rates of inflation over several years: 'The guesswork and the gambling [on prices] left a lot of sleepless nights.'[30] The quarrying side of the business was not so badly affected by the escalation in prices, as road surfacing contracts were re-negotiated on an annual basis with local authorities. But the inflationary economic environment reduced overall profit margins and greatly diminished

29. Interview with Niall Collen, 14 May 2009.

30. Interview with Niall Collen, 5 February 2010.

the returns that the firm offered to the family. The commercial pressures generated by inflation contributed to a fundamental reappraisal of the company's operations and profile during the 1970s.

It is likely that a reassessment of the firm's direction would have occurred around this time anyway, as the family proprietors were approaching retirement. Joe and Tony Collen had directed the company north of the border, in various incarnations, for almost fifty years. They had been responsible for the fortunes of the family business through economic depression, global conflict and post-war expansion. The two men were ready to hand over responsibility to the next generation by the late 1970s. Joe Collen told the Rotary Society in 1971 that he had found 'plenty of excitement' in the building trade but also noted wryly that 'pricing bills of quantities … wondering if the price is correct for the job, it is one of the best trades for creating ulcers.'[31] The commercial pressures intensified throughout the 1970s, not least due to the corrosive impact of inflation on the building industry. The turbulent economic conditions led to internal deliberations among family members about the future of the business during the mid-1970s. The company was able to employ a substantial workforce of about 300 and was not a loss-making concern, but there was a feeling among family members that the firm was no longer commercially successful and was unable to provide a reasonable livelihood for the directors. Niall Collen summed up the concerns of the senior proprietors: 'The fun had gone out of the business for them [Joe and Tony]. The profit margins in contracting weren't great at that stage. While we weren't losing money, we weren't making a lot … the family generally felt that there wasn't the profit in it to justify carrying it on in its existing form.'[32] The deliberations within the family produced a consensus that their existing business model was no longer sustainable by the late 1970s.

The family proprietors adopted a course of action that decisively reshaped the firm, deciding that Collen Brothers should be wound up and replaced with a new, smaller company, which would concentrate on the quarrying side of the business. A quarrying enterprise appeared a much safer and potentially more profitable undertaking than building, due to the regular nature of road surfacing work with the local authorities and the firm's ability to control costs more tightly on one-year contracts. The decision was linked to a generational transition within the firm. Niall Collen, who was a leading participant in

31. Speech by Joe Collen, *My Job*, March 1971, p.3.
32. Interview with Niall Collen, 14 May 2009.

the deliberations which reshaped the company, was particularly interested in developing the quarrying side of the business.[33] Yet the decision itself was undoubtedly dictated by commercial considerations; major construction projects offered considerable risk and little reward in the fraught economic environment of the time.

The transformation of the firm was accomplished in the late 1970s. Collen Brothers Ltd was placed in voluntary liquidation in 1979, although the original company remained formally in existence for over a decade.[34] Tony and Joe Collen had established a subsidiary company in December 1958, which was used for various commercial purposes up to the 1970s. In a fine example of corporate recycling, the family used it as a vehicle to establish their new company, under a title that illustrated the changing character of the business. Collen Brothers (Quarries) Ltd was formally inaugurated on 7 February 1978.[35] The new company adopted a memorandum of association which described its commercial mission: 'To carry on all or any of the business of manufacturers of and dealers and workers in aggregate and bituminous products … and to buy, sell and deal, either wholesale or retail, in the above products, and [act as] quarry owners, builders, general contractors …'[36]

While building was included as a possible activity of the company, its founding charter placed an unmistakable emphasis on the management of the quarry and the sale or commercial use of its products. Moreover, it soon became apparent that the name of the company was the most reliable indication of its core activities. The new company focused on the development of the quarry, the manufacture of bituminous products and the expansion of civil engineering work associated with quarrying, notably the road-surfacing and maintenance projects. Many of the traditional features of the business disappeared during this period. The joinery department was closed in the mid-1970s, foreshadowing the full-scale restructuring of the company itself towards the end of the decade. The building business was gradually but completely run down, while the firm also ceased to undertake large-scale civil engineering jobs.[37]

---

33. Interview with Niall Collen, 14 May 2009.

34. Interview with Niall Collen, 5 February 2010.

35. Northern Ireland Companies Office, NI 4186, *Certificate of Incorporation on Change of name*, 7 February 1978.

36. Collen Papers Portadown, Minute book, Collen Brothers (Quarries) Limited, Minutes, Annual General Meeting, 5 October 1979.

37. Interview with Niall Collen, 5 February 2010.

'After a quick one' – Collen employees emerging from a pub in Bellaghy.
Courtesy of Eleanor Gough (née Collen).

The decisions taken by the proprietors brought a significant contraction in the turnover of the company. Collen Brothers Quarries was a much smaller business than its predecessor – the workforce was gradually reduced during the 1970s, so that the new company retained only forty employees by the end of the decade.[38] A combination of commercial pressures and the retirement of the senior proprietors generated a sweeping transformation of the company.

The establishment of the quarrying company also brought a swift and amicable transfer of authority to a new generation. Niall Collen took over the management of the business, taking up office as chairman and managing director at the first board meeting of the new company on 27 February 1979.[39] The shares in Collen Brothers Quarries were distributed on a 50:50 basis between Niall Collen and Tony Collen's three children, Eleanor Jackson, Noranne Biddulph and Frances Bolger.[40] The composition of the new board reflected the equal split in the shareholding. The other directors appointed in February 1979 included Mary Collen, wife of Niall, who acted as company secretary, Eleanor Jackson and Noranne Biddulph.[41] The far-reaching changes

38. Interview with Niall Collen, 14 May 2009.

39. Collen Papers Portadown, Minute book, Collen Brothers (Quarries) Limited, Minutes, first Meeting of Directors, 27 February 1979.

40. Minute book, Collen Brothers (Quarries) Limited, Minutes, Annual General Meeting, 2 July 1980. Eleanor Collen has married twice, first to Len Jackson and later to Johnny Gough.

41. Minute book, Collen Brothers (Quarries) Limited, Minutes, first Meeting of Directors, 27 February 1979.

in the character and management of the firm were accomplished with the full agreement of the senior family members, who did not simply fade into the background. Tony and Joe Collen were appointed as consultants to Collen Brothers Quarries: both men remained regularly in attendance at general meetings of the company into the twenty-first century, displaying a remarkable record of longevity even by the impressive standard established by their predecessors. The new company remained very much a family business. Yet in other ways the firm had broken with the past in a dramatic fashion. The Collen family had been engaged in the building trade long before the emergence of the company itself, while building operations had been at the heart of Collen Brothers since its foundation. The emergence of the new company marked the first time in over 150 years that no member of the Collen family was involved in building work in Portadown. The decisions taken in this period reflected the unsentimental pragmatism of the proprietors and their keen awareness of commercial realities – ultimately the directors, both new and old, set out to guarantee the survival of the family business.

THE TROUBLES: 'WE WOULD HAVE WORKED ANYWHERE, GREEN OR ORANGE WOULDN'T MATTER TO US'

The company had operated since its foundation in a society which was divided by conflicting religious and national allegiances. Following the partition of the island and the establishment of the devolved administration in Northern Ireland in 1921–22, a deeply entrenched community division between the unionist majority, represented by the Stormont administration, and the nationalist minority became a central feature of the politics and society of Northern Ireland. Intermittent outbreaks of violence orchestrated by the illegal Irish Republican Army (IRA), though entirely unsuccessful in achieving the movement's declared objective of a united Ireland, inflamed the suspicions of unionist politicians against the minority and served as a further justification for one-party rule by the Unionist Party at Stormont. The emergence of a highly effective and vocal civil rights movement, demanding far-reaching reforms to guarantee equality for the Catholic minority, closely followed by a loyalist backlash against any real reform, provoked a crisis within the Northern Ireland state in the late 1960s.[42] The outbreak of clashes between nationalist

---

42. Joseph Lee, *Ireland 1912–85:Politics and Society* (Cambridge, 1989), pp.416–20.

demonstrators and the security forces in Derry in August 1969 served as the catalyst for the eruption of widespread violence, which soon extended to other towns and cities of Northern Ireland.[43] The Provisional IRA began a bombing campaign in 1971, triggering a sustained conflict between militant nationalism and the British state. The imposition of direct rule from Westminster in March 1972 could not mitigate the increasingly endemic violence by both republican and loyalist paramilitary groups. The conflict settled into a grim stalemate by the mid-1970s, in which the IRA failed to force the withdrawal of the British forces, but nor could the security forces defeat the IRA militarily. This political and military deadlock was maintained until the 1990s, when an inclusive peace process brought an end to the IRA campaign and led to a historic power-sharing agreement in 1998. The Belfast Agreement itself attracted intense controversy, but served as the basis for a historic accord that culminated in the formation of a power-sharing administration including the major unionist and nationalist parties in 2007.

The Troubles formed a dismal backdrop to Collen's activity in Northern Ireland for a quarter of a century. The company had always avoided political or religious considerations in pursuing its business; its reputation as an established local firm, which had carried out building projects for all the major denominations in Northern Ireland, proved a valuable advantage during the conflict. Collen did not differentiate between employees of different religious traditions, employing managers and workers without regard to their religion.[44] Niall Collen recalled that 'we would have worked anywhere, Green or Orange wouldn't matter to us'.[45] The family proprietors acted to minimize the risks to their employees and safeguard the company itself. Collen had undertaken work for the Royal Ulster Constabulary (RUC) before the Troubles, although it was never a major element of the company's business. The directors took a deliberate decision not to undertake work for the security forces throughout the conflict. While other construction companies secured valuable contracts from the army or police, Collen chose not to seek such business, which had the potential to turn the firm and its employees into targets of the IRA campaign.[46] This practice applied to the original company from the early 1970s and to its successor, Collen Brothers (Quarries). The family was concerned to maintain

---

43. Justin O'Brien, *The Arms Trial* (Dublin, 2000), p.35.

44. Interview with Niall Collen, 14 May 2009.

45. Interview with Niall Collen, 5 February 2010.

46. Interview with Niall Collen, 14 May 2009.

a low-key method of operation and to avoid any strong political or religious identification associated with the Troubles. Their cautious and pragmatic approach was underlined by an incident in the mid-1970s, when the RUC offered to issue the directors with private firearms for their own protection, considering that family members might be potential targets due to Collen's work for the police in the past: a publication by a postgraduate at Queen's University had included the company on a list of construction firms who had previously carried out work for the security forces. The directors politely rejected the offer, as Niall Collen noted: 'No, at that stage we wanted to play the whole thing down and this wasn't our scene. We were quite happy not to be seen as particularly strong loyalists or whatever. We walked the middle path.'[47] The directors' approach was certainly effective judged by the most essential consideration, as none of the firm's employees lost their lives during the conflict. The company also managed to protect its premises and operations from any deliberate targeting or terrorist attack. The firm's pragmatism, low-key *modus operandi* and commercial ecumenism served it well throughout the Troubles.

Yet the firm could not entirely escape the pervasive nature of the conflict. Collen came under fire – sometimes literally – from both sides of the sectarian divide at different times, fortunately without suffering any loss of life. The company's attitude during the Loyalist Workers' Strike, which aimed to overthrow the fragile power-sharing executive set up following the Sunningdale agreement in 1973, illustrated the determination of the directors to remain aloof from the conflict. The strike, which was supported and in some areas actively enforced by loyalist paramilitaries, brought down the short-lived power-sharing experiment in 1974.[48] The Loyalist Workers' action was strongly supported in Portadown and the surrounding areas, which remained a bastion of Unionism and of the Orange Order. Yet the senior proprietors, who were still directing the company's operations, declined to participate in the strike. The company's stand drew the anger of the strike organisers, as a long-time company insider noted: 'It was decreed that businesses should close and so on and we didn't and of course there were road blocks. Then the next day we came back to the quarry premises, to find that hydraulic pipes had been cut in the mobile machinery.'[49] The vandalism of equipment at the firm's premises reflected the difficulties

---

47. Interview with Niall Collen, 14 May 2009.

48. Ferriter, *Transformation of Ireland*, p.629.

49. Interview with Niall Collen, 14 May 2009.

faced by the directors in maintaining 'the middle path', but it did not deter them from their chosen course of action.

A more dramatic and dangerous incident occurred in Belfast, also in the early 1970s. The company used a tanker to transport liquid bitumen from Belfast to their Tandragee quarry. One of its drivers was in Belfast to collect a load when he had a narrow escape as the IRA attacked a British army patrol on the M1. Niall Collen retained a vivid recollection of the life-threatening incident: 'One of our drivers got caught in the gunfire from an IRA ambush, set up near Casement Park, of a British army patrol on the M1 motorway. Thankfully he wasn't hit – the bullet holes were there to be seen in the cab.'[50]

The driver survived unscathed and remained an employee of the company. The IRA ambush was not the only occasion on which the company became caught up inadvertently in the conflict. On another occasion in 1974, one of the company's JCB diggers was seized by loyalist rioters in Lurgan and used as part of a barricade against the security forces.[51] Shortly afterwards when the disorder had abated, Niall Collen went to the town with two company employees to salvage the digger. Although its tyres had been slashed to prevent it being moved, they managed to retrieve the machine and bring it back to Hanover Street, where it was soon pressed back into service.

Perhaps the most striking aspect of Collen's operations during the Troubles is not that such moments of danger or disruption occurred, but that so few arose in such a prolonged conflict. The directors and employees undoubtedly took seriously the various threats presented by the Troubles, avoiding contracts that might have attracted hostile attention and not taking work in places which were considered particularly dangerous. But they also regarded the occasional moments of danger as isolated incidents rather than a normal pattern of life. The experience of the Collen family and their employees appeared to demonstrate that it was possible to live and work normally, in a region wracked by conflict, to an extent that was surprising to outside observers. Yet Collen's ability to operate without crippling problems to its operations or still worse, personal tragedy among its staff, was not simply a matter of good fortune, but owed much to the character of the company itself and the quality of its leadership. The directors were remarkably successful in minimizing the impact of the Troubles on the family business.

---

50. Interview with Niall Collen, 14 May 2009. 'Eleanor Collen has married twice, first to Len Jackson and later to Johnny Gough.

51. Interview with Niall Collen, 5 February 2010.

Surface dressing, 1981; George Deens is the driver of the tanker.
Courtesy of Peter Collen.

COLLEN BROTHERS (QUARRIES)

The company was also obliged to contend with an unfavourable economic environment during the 1980s, due to the continuing prevalence of inflation and an international recession early in the decade. Collen Brothers (Quarries) relied heavily on public contracts for road surfacing and maintenance. Armagh County Council was initially the major client for the new company, until the direct rule administration took charge of the delivery of public works from the local authorities. The administrative restructuring, which gave the Roads Service within the Department of the Environment responsibility for road maintenance, did not affect the company's business in the short term, and Collen secured annual contracts with the Roads Service every year from its inception until 2004. The company mainly worked for the southern division of the Roads Service, which had responsibility for the upkeep of roads in Co. Armagh, but also undertook road surfacing contracts in other areas of Northern Ireland.[52] The firm developed additional road surfacing capabilities in the early 1980s: Collen took on 'surface dressing' contracts for the Roads

---

52. Interview with Niall Collen, 5 February 2010.

Service, utilising aggregate produced in their own quarry.[53] Surface dressing is a relatively cheap method of repairing or resurfacing roads, involving the spraying of bitumen emulsion, consisting of a mixture of bitumen and water, on the road, followed by the spreading of chippings held in place by the bitumen. The company also became involved in the production and laying of asphalt, a denser and more durable bituminous mix, which was used especially for the surfacing of major roads. Road surfacing was the core of the business, reflecting the company's concentration on developing the quarry and making full use of its materials.

The company was arguably less severely affected by the Troubles than by the economic policies of the British government in the 1980s. While the British state subsidized the Northern Ireland economy throughout the Troubles, the region did not escape the impact of public spending cuts imposed by Margaret Thatcher's government following her election in 1979. The new company suffered from a retrenchment in the road-maintenance business as state funding for public works was reduced. The minutes of the second AGM of Collen Brothers Quarries on 25 September 1981 noted that 'the company was heavily dependent on government expenditure for the maintenance of roads and this was declining fairly rapidly'.[54] The company's turnover and profits declined in the early 1980s, reflecting adverse trading conditions and the tighter constraints on public expenditure enforced by the Conservative government. The firm's activity revived towards the middle of the decade in line with a wider recovery in the economy, as its turnover increased by 10 per cent in 1984. Yet the pace of economic growth was intermittent, contributing to uncertain trading conditions for the company. A major upturn in the construction industry in England during the mid to late 1980s did not reach Northern Ireland.[55] The firm's turnover remained under £1 million until 1987, while a moderate increase in sales in the late 1980s did not translate into healthy profit figures.[56] The competition for public contracts was intense; Niall Collen informed the AGM in April 1989 that the firm had lost out on several major road surfacing contracts due to the willingness of several competitors to work at rates that he considered 'uneconomic'.[57] The company,

---

53. Correspondence with Niall Collen, 16 March 2009.

54. Collen Papers Portadown, Minute book, Collen Brothers (Quarries) Limited, Minutes, Annual General Meeting, 25 September 1981.

55. Minute book, Collen Brothers (Quarries) Limited, Minutes, AGM, 5 May 1988, p.1.

56. *Ibid.*

57. Minute book, Collen Brothers (Quarries) Limited, Minutes, AGM, 17 April 1989, p.1.

however, generally held its own in competitive tendering, acquiring its fair share of road-surfacing contracts from the Department of the Environment, but was obliged to tender for jobs at highly competitive prices that left very tight margins for profit. The managing director's report to the AGM in May 1990 noted that during the previous year 'there was a lot of work for very little profit'.[58] The company maintained a considerable level of activity, but its profits remained modest as a proportion of turnover for most of the 1980s.

Collen achieved a rapid expansion in its turnover in the early 1990s. The firm's turnover increased steadily at first, from £1.5 million in 1989 to £1.8 million by 1993; then came a more dramatic advance to £3 million in 1994. The abruptness of the latest increase was due in part to the company's success in winning a major contract for road surfacing on the outskirts of Belfast.[59] But the expansion in turnover was not simply a once-off development, as it also reflected a steady increase in the sales of more valuable and profitable products, such as asphalt.[60] Profits also showed a conspicuous upward trend for much of the 1990s; the firm reported an increase of 25 per cent in gross profit for 1993 alone.[61] But results later in the decade were less impressive, as the construction industry experienced one of its many cyclical troughs towards the end of the 1990s. The election of Tony Blair's New Labour government in 1997 had little short-term impact on the economic situation in the region. The new government initially showed considerable caution about the control of public spending, seeking to establish its credentials for economic management. The managing director's report to the company's AGM in 1998 noted with disapproval that 'the Labour government have continued with the Tory policy of reducing public spending.'[62] Yet despite such unwelcome continuity in government policy, both turnover and profit remained at a relatively high level, especially in comparison to the harsher economic conditions of the previous decade. The company reported turnover of just under £3 million and a gross profit of 17.9 per cent in 1999.[63] Collen Brothers Quarries was a successful company with a healthy financial position and an established niche in the marketplace at the end of the twentieth century.

---

58. Minute book, Collen Brothers (Quarries) Limited, Minutes, AGM, 15 May 1990, p.1.

59. Minute book, Collen Brothers (Quarries) Limited, Minutes, AGM, 15 May 1995, p.1.

60. *Ibid.* p.2.

61. Minute book, Collen Brothers (Quarries) Limited, Minutes, AGM, 9 May 1994, p.1.

62. Minute book, Collen Brothers (Quarries) Limited, Minutes, AGM, 3 April 1998, p.2.

63. Minute book, Collen Brothers (Quarries) Limited, Minutes, AGM, 18 April 2000, p.1.

Collen Brothers' quarry at Tandragee, 1984. Courtesy of Peter Collen.

The board of directors under Niall Collen's leadership took several initiatives in the final decades of the century that significantly enhanced the competitive position of the company. The board initiated the construction of a new asphalt manufacturing plant in 1985, replacing the existing facility. This investment gave the company the benefits of a modern plant for the manufacture of a high-quality product.[64] The ability to manufacture asphalt gave Collen a competitive edge over many of its rivals, allowing the firm to control its costs and achieve a higher profit margin. Following the construction of the new plant, the company steadily expanded its sales of asphalt over the following decade.[65] Niall Collen also initiated a gradual modernization of the firm's equipment and machinery. The company invested in new primary, secondary and tertiary crushers for quarry work, replacing older machines that had become obsolete.[66] The directors also authorized investment in replacement equipment for road surfacing projects, including Hamm and Phoenix chipping machinery for surface dressing and new Blawknox and ABG pavers for laying bitmac and

64. Minute book, Collen Brothers (Quarries) Limited, Minutes, AGM, 13 August 1985, pp.1–2.

65. Minute book, Collen Brothers (Quarries) Limited, Minutes, AGM, 5 May 1988, p.1; Minute book, Collen Brothers (Quarries) Limited, Minutes, AGM, 9 May 1994, p.1.

66. Minute book, Collen Brothers (Quarries) Limited, Minutes, AGM, 30 April 1993, p.3; Interview with Niall Collen, 5 February 2010.

A CAT 966F loading shovel, on display at the Balmoral Show, 1994.
Courtesy of Peter Collen.

asphalt.[67] But the most significant advance for the long-term security of the company was the acquisition of the quarry, which was the essential hub of the business from the late 1970s.

Niall Collen identified the expansion and development of the quarry as a key priority immediately after the establishment of the new company. The managing director successfully renegotiated the lease with the Duke of Manchester in 1980, to provide for additional land at Tandragee for quarrying. The company also secured the option to continue quarrying rock under this land for up to thirty years from 1 January 1979.[68] This proved to be only an interim arrangement, however, as the firm moved to buy the land itself a decade later. The directors agreed in April 1989 to initiate negotiations with the Duke's agents to purchase the quarry 'if the price was not prohibitive...': Niall Collen hoped to secure ownership of the site before undertaking a more intensive development of the quarry, including a deeper excavation of the quarry floor.[69] The discussions continued for over a year, in part because the Duke was initially reluctant to sell the land, preferring a regular income to a

67. Minute book, Collen Brothers (Quarries) Limited, Minutes, AGM, 18 March 1996, p.2.

68. Minute book, Collen Brothers (Quarries) Limited, Minutes, Meeting of Directors, 7 May 1980.

69. Minute book, Collen Brothers (Quarries) Limited, Minutes, AGM, 17 April 1989, p.3.

lump sum payment. When he agreed to consider a sale, his representatives raised the possibility of selling the quarry instead to the local district council.[70] The company's representatives pointed out, however, that the lease held by the firm did not allow the proprietor to sell to another party, removing the unwelcome prospect that the most essential element of the company's business might come under the control of another and possibly more unpredictable landlord. The negotiations produced a settlement acceptable to the Duke in 1990, involving the purchase of the quarry by Collen at a cost of £120,000. When the negotiations with the Duke's agents were completed, the directors and shareholders held an extraordinary general meeting on 5 December 1990 to finalize the agreement. Tony and Joe Collen attended the meeting, reflecting their status as the elder statesmen of the firm. Although most directors were concerned about the potential downside of owning the site after quarrying ceased, they reached a consensus to proceed with the purchase of the quarry.[71] Niall Collen was undoubtedly the driving force behind the acquisition of the quarry, while the other directors also recognized the importance of owning the firm's most significant asset.

The company began to develop a more diverse client base during the 1990s. While Collen continued to secure public contracts, the proportion of work undertaken for the Roads Service declined as the company acquired more work from private clients.[72] The company supplied quarry products or undertook road surfacing for industrial and residential developers as well as public bodies. It was fortunate that Collen had moved away from its traditional dependence on public contracts, as the firm lost out in new contract arrangements adopted by its most established client early in the twenty-first century. The Department of the Environment in 2004 introduced new long-term contracts for each division of the Roads Service, amalgamating the wide range of annual contracts for road surfacing and maintenance into a much larger package spread over five years. The directors were concerned about the proposal as soon as it emerged – the company's AGM in 2002 noted that the reorganization might place Collen, as a relatively small company, at a disadvantage in tendering for the new contracts.[73] This assessment proved entirely

70. Minute book, Collen Brothers (Quarries) Limited, Minutes, AGM, 15 May 1990, p.3.

71. Minute book, Collen Brothers (Quarries) Limited, Minutes, Extraordinary general meeting, pp.1–4; *Memo from Chairman to Secretary*, 5 December 1990.

72. Minute book, Collen Brothers (Quarries) Limited, Minutes, AGM, 18 March 1996, p.3.

73. Minute book, Collen Brothers (Quarries) Limited, Minutes, AGM, 30 April 2002, p.3.

accurate. Collen lost out in 2004 for the two major contracts awarded by the southern division of the Roads Service under the new system. The company failed to secure a five-year contract for surface dressing in its normal area of operation and later in the same year, Collen formed part of a consortium of four companies of similar size that came second in the tendering process for bitumen macadam and asphalt surfacing for the same area.[74] The loss of such valuable long-term contracts, involving projects that had traditionally been the mainstay of the business, was a significant setback. But the impact on the firm was much less than it might have been a decade earlier. Niall Collen's report to the AGM in 2004, before the outcome of the second tender was known, noted that 'our company can survive happily with private work if we should not be successful.'[75] The company adapted effectively to the new commercial situation by acquiring a higher volume of private work. Collen undertook road surfacing work for factory car parks, schools and commercial developments; the company also surfaced ancillary roads in new private developments.[76] The firm benefited from an upsurge in the property market in the region during the first decade of the new century. Collen, however, did not abandon its traditional sources of turnover, continuing to undertake road maintenance contracts for public bodies where such work remained a practical possibility. Indeed the company secured annual contracts for surface dressing in the Newry and Mourne area, where the new format had not yet been rolled out, around the same time that it lost out in the allocation of the other projects for the southern division.[77] Yet the balance of the company's activity shifted decisively to the private sector, marking a striking change in the traditional profile of its quarry operations since the Second World War.

The first decade of the new millennium brought the involvement of a seventh generation of the Collen family in the business. Two of Niall Collen's three sons took up managerial responsibilities within the firm. Andrew, a graduate of Heriot-Watt university in Edinburgh, worked initially with Tarmac in its quarry operations in Britain, before returning home to become assistant manager of the quarry at Tandragee in 2004 – he took over as quarry manager

---

74. Minute book, Collen Brothers (Quarries) Limited, Minutes, AGM, 29 April 2004, p.2; Minutes, AGM, 27 April 2005, p.2.

75. Minute book, Collen Brothers (Quarries) Limited, Minutes, AGM, 29 April 2004, p.3.

76. Interview with Niall Collen, 5 February 2010.

77. Minute book, Collen Brothers (Quarries) Limited, Minutes, AGM, 27 April 2005, p.2.

in April 2006.[78] More recently, Peter, having graduated from Trinity College and the Smurfit business school in UCD, also joined the company, becoming involved in managing its head office in Portadown.

Collen and Carson: the company at work at Stormont, under the watchful eye of Lord Carson, January 2009. Courtesy of Peter Collen.

Collen in Northern Ireland charted its own distinctive path, influenced by the political and economic environment in which it operated. The firm remained a family business in the most classical form: the company was not only owned and managed by the family but its operations were overseen on a day-to-day basis by family members. This method of operation remained the

---

78. *Ibid.*; Minutes, AGM, 10 May 2006, p.3.

consistent organizing principle of the firm throughout two generations and was the most striking element of continuity at a time of far-reaching change within the company itself. The firm in Portadown adapted with impressive skill and resilience to the extraordinary circumstances of the Troubles, but its development was never dictated by political or cultural imperatives. It was the prevailing economic conditions, combined with a generational transition, which provoked a transformation within the firm during the 1970s, leading to a refocusing of its commercial activity and a significant break with traditional elements of its business. The establishment of Collen Brothers (Quarries) entailed a far-reaching rationalization of the firm, but also created a more financially sustainable and profitable business. The new company was very different from its southern counterpart and indeed from the traditional firm in Portadown itself a generation previously, but it proved sufficiently flexible and durable to prosper in an era of political and economic upheaval.

SEVEN

Embracing the Celtic Tiger?

The company in Dublin suffered similar pressures to its northern counterpart in adapting to the volatile economic conditions created by the second oil crisis and the ensuing international recession, but it benefited from a much more dramatic upsurge in economic activity during the late 1990s. The transfer of ownership within the company in 1996 coincided with the emergence of an economic boom of unprecedented scope and duration, which would last for over a decade. Collen adapted its business model to take full advantage of an era of extraordinary economic expansion. The company under Neil Collen's direction undoubtedly took initiatives that would not have occurred in the previous generation, notably a sweeping redevelopment of the company's premises at East Wall and a wider engagement with the development of commercial and residential property. The change in ownership paved the way for significant changes in the firm, which were soon visible in a radical restructuring of its commercial organization and in a remodelling of its physical infrastructure. But neither the Celtic Tiger nor generational transition led the company to join in the reckless exuberance displayed by some of its contemporaries at the peak of the economic boom. The company certainly prospered during the Celtic Tiger, but did not place short-term profit before the long-term viability of its business.

Neil Collen initiated a sweeping reorganization of the company within a year of his appointment as managing director. The restructuring was intended primarily to facilitate independent activity by the design function, recognizing its increasing contribution to the company's turnover. Collen had operated since the 1960s as a building company with a strong internal design function that devoted its energies exclusively to design and build contracts for the company itself. The design office expanded its traditional role from the late 1980s, managing the design side of projects for external clients.[1] One of the first projects managed by the office was the design of a new sports hall for St Columba's College in south Dublin. The design office gradually evolved into a project management division with a growing client list by the mid-1990s. The minutes of Collen Construction recorded an extensive volume of activity by the project management division in 1997, including the design of a new stand for Fairyhouse racecourse, design work with SIAC Construction on a new building for Tobacco Distributors and the planning of a new sports hall for Castlepark school.[2] It was particularly striking that the division was working in conjunction with SIAC, a major building and civil engineering contractor that had previously been in competition with Collen Construction. The design function was increasingly charting its own distinctive path within the company during the 1990s, securing valuable contracts and forging commercial alliances with other building contractors.

The formal structure of the company, however, still reflected the practices of a very different era. Although the design office was incorporated as a subsidiary company as early as 1970, in effect it remained a trade division of Collen Construction until the late 1990s. The expansion of the design company's remit was acknowledged in January 1994 when it was given a new trade name as Collen Project Management, but this re-designation was not yet accompanied by formal autonomy for the newly renamed division, and the directors of Collen Construction continued to control the subsidiary company.[3] This structure lacked flexibility and complicated the company's ability to tender successfully for contracts that did not fall within its favoured

1. Interview with Chris Lyons, 6 July 2009.

2. Collen Construction, Minutes, Meeting of Directors, 16 April 1997, p.2.

3. Collen Project Management Ltd, Minutes, Final meeting of directors of Collen Design, 27 January 1994.

design and build model. Neil Collen recognized the limitations of the traditional structure:

> Design and build is a very good way [of operating] but in Ireland particularly it has never been embraced as much as in other places. So sometimes … if a contractor has got a design capability [as Collen does] when you're going to tender and you're looking to architects to give you opportunities they may feel that you're competing against them directly.[4]

The existing structure limited Collen Construction's potential for expansion in straightforward building contracts, while also restricting the project management division's ability to attract outside business. Neil Collen therefore acted to develop a group structure that would give autonomy to the design and construction divisions of the firm while maintaining oversight through a new group holding company.

The plans for reorganization were first raised at a board meeting of Collen Construction on 15 December 1996, when the directors agreed to ask the Dublin branch of Ernst & Young, the international firm of business advisers and auditors, to advise on the implementation of the scheme.[5] The board moved decisively to implement the new group structure, giving the green light to a reorganization plan produced by Ernst & Young on 14 May 1997.[6] A month later on 18 June the directors agreed to 'request Ernst & Young to proceed with the scheme as speedily as possible.'[7] The initiative was implemented in a series of stages, beginning with the establishment of a new holding company, Collen Group Ltd, which was incorporated on 1 August 1997 to provide the institutional framework for the new structure. The newly constituted company remained firmly under the ownership and control of the family; initially the directors were Neil Collen, his wife Pamela and Lyal. The older man had served on the board of the firm throughout its various incarnations since 1949. Neil and Pamela Collen were the main shareholders, while Lyal also retained a modest stake in the business.[8] The new company took over the ownership of Collen Construction and also controlled any new subsidiary companies set up as part of the reorganization.[9] The restructuring

---

4. Interview with Neil Collen, 28 April 2009.

5. Collen Construction, Minutes, Meeting of Directors, 15 December 1996, p.2.

6. Minutes, Meeting of Directors, 14 May 1997, p.2.

7. Minutes, Meeting of Directors, 18 June 1997, p.1.

8. Collen Group Ltd, Minutes, Meeting of Directors, 12 November 1997.

9. Collen Construction, Minutes, Meeting of Directors, 13 November 1997, p.1; Minutes, Meeting of Directors, 18 November 1997, p.1.

was designed to facilitate the establishment of autonomous companies with distinct functional responsibilities under the auspices of Collen Group.

A gradual process of decentralization unfolded over the following two years. A new subsidiary company, Collen Properties, was established early in 1998, taking ownership of the company's newly built headquarters at River House.[10] Collen Properties was an investment company that was originally intended to serve as a vehicle for future property investments, but the board soon determined that any new property would be vested in its own subsidiary company – River House remained the sole asset of Collen Properties. But the most significant innovation flowing from the restructuring process was the emergence of distinct companies dedicated to building and project management. Neil Collen noted: 'I was trying to create a structure where the various elements were concentrating on their core abilities and experience, the designers were designing, the builders were building … and we also have the group which undertakes developments for ourselves.'[11]

The managerial responsibility for design and construction was more sharply defined within Collen Construction in June 1997, when the board delegated the running of the project management division to Chris Lyons and Pat Sides, while Martin Glynn took overall charge of the construction side of the business.[12] This was only an interim arrangement. Collen Project Management (CPM) was established as a separate company with its own board of directors on 1 April 1999, allowing the design function to operate in its own right for the first time.[13] Collen Construction became purely a building company, retaining responsibility for the original core business of the firm.

The reorganization of the company did not dilute the central position of the Collen family; indeed Collen Group was the latest incarnation of the family business. But in other respects the restructuring delivered far-reaching changes in the way in which the firm had operated for over a century. Whatever its corporate label, the company had usually functioned as a centralized entity under close supervision by one or more family members since its foundation. The group structure involved an unprecedented level of delegation

10. Collen Construction, Minutes, Meeting of Directors, 25 February 1998, p.2; Interview with Paddy Wall, 28 April 2009.

11. Interview with Neil Collen, 28 April 2009.

12. Collen Construction, Minutes, Meeting of Directors, 18 June 1997, p.1; Minutes, 10 December 1997, p.2.

13. Minutes, Meeting of Directors, 23 February 1999, p.2.

on the part of the family directors. This did not mean that family members had abdicated their traditional role in managing the business: Neil Collen became managing director of the Group and acted as chair of all the subsidiary companies from 1999. But the reorganization brought a formal devolution of authority to professional managers that had never occurred before. The new structure required autonomous boards for each company, which were composed predominantly of senior managers and other experienced employees drawn from outside the family. Martin Glynn was appointed as managing director of Collen Construction on 26 April 1999, while two senior quantity surveyors, Alan Brunton and Declan Lowry, were also elevated to the board at the same time.[14] Leo Crehan, who had previously worked with Collen Brothers from 1978 to 1982, returned to the building company as contracts director in May 2000.[15] Chris Lyons, the chief architect and longest serving member of the design office, became managing director of CPM; four other directors were appointed to the board of the company, including experienced employees such as Pat Sides and Frank O'Sullivan.[16] The managers of the subsidiary companies were given considerable autonomy in running their section of the business, within the framework of broad policy decisions set by Collen Group.[17] Moreover this process of delegation extended to the holding company itself. Paddy Wall, the company secretary since 1983, was appointed as a director of Collen Group in April 1999, at the same time as the subsidiary companies were allowed to develop autonomously.[18] Neil Collen showed a much greater willingness to delegate authority than most of his predecessors, especially Standish, the dominant figure in the management of the firm during the previous generation. While Collen remained very much a family business, the reorganization enshrined institutional autonomy as a central element of the managerial structure. The restructuring acknowledged the reality that the company could no longer be effectively overseen by a single individual or even a small core management team under tight central control.

The reorganization of the company occurred in the context of a rapid expansion in the construction industry and the wider economy. The export-led economic development of the late 1990s produced impressively high

---

14. Minutes, Meeting of Directors, 26 April 1999, p.1.

15. Minutes, Meeting of Directors, 1 June 2000, p.1.

16. Collen Project Management, Minutes, Meeting of Directors, 1 April 1999, p.1.

17. Interview with Paddy Wall, 28 April 2009.

18. Collen Group, Minutes, Meeting of Directors, 1 April 1999.

growth rates and a dramatic expansion in employment, with the creation of 600,000 jobs in a ten-year period between 1997 and 2007.[19] The Celtic Tiger, as its political and media admirers dubbed the Irish economic transformation, undoubtedly marked an impressive achievement for a state, which had traditionally lagged behind most of its European neighbours. Collen flourished in the newly benign economic environment. The company enjoyed an unprecedented expansion around the beginning of the new millennium, fuelled by a high demand for construction projects and by a booming property market. The turnover of Collen Group increased more than six fold in less than a decade, from £15,648,328 in 1998–99, the first full year of its operation, to just under €100 million by March 2007.[20] The effect of the economic boom on the profit margins was even more striking. The company had barely struggled back into the black in 1994–5 and returned distinctly modest profits during the mid-1990s.[21] Shortly after its foundation, Collen Group recorded a respectable profit before tax of £584,215 in 1998–9, which was not dissimilar to the company's performance in previous periods of economic buoyancy. But the impact of the Celtic Tiger was very different – eight years later the group reported a phenomenal profit of almost €9 million, the largest ever achieved by the firm. The company enjoyed an extraordinary degree of commercial success, which the directors could not have contemplated even a decade earlier. Collen's expansion owed much to the rising economic tide, which created an unusually wide range of opportunities for its construction, design and investment companies. But the company's success was also facilitated by the way in which the directors put their own house in order, completing the process of internal succession and implementing a far-reaching restructuring of the firm.

The most visible result of the change in ownership within the company was an extensive programme of redevelopment at its premises on East Wall Road. The development of the property followed a gradual decline of the traditional operations in the builders' yard. Much of the work traditionally carried out by employees in the yard had been taken over by subcontractors over the previous decade. The activity of the joinery shop and fitters' shop had already declined during the recession in the 1980s. Both trades were

---

19. Murphy, *Promised Land*, p.14.

20. Collen Group, Accounts, *Consolidated Profit and Loss Accounts for the year ended 31 March 2000*, p.7; Collen Group, Accounts, *Consolidated Profit and Loss Accounts for the year ended 31 March 2007*, p.6.

21. Collen Construction Ltd, Accounts, *Profit and Loss Account for the year ended 31 March 1995*, p.6.

profoundly affected by the move from direct labour to subcontracting. The fitters' shop was the first to close, while the operations in the joinery shop were winding down by the early 1990s. While the joinery shop continued to produce high-quality products, its work could be done at a much more economical rate through subcontracting. The joinery shop was closed in 1995, as a result of the board's decision to proceed with the redevelopment of the yard. Jimmy Small, who headed the joinery shop from the 1970s until its closure, noted that changing economic and technological trends drove the decline of the traditional craft operations in the yard:

> So things were changing slowly, whereas you would give a price for joinery and you would hope that you could make it for that, you could now get a price from a joinery shop for making it for you, put it onto the bill of quantities and away you went. So in a sense it was a comfortable way of doing things and you didn't have problems like an accident or a machine breaking down or whatever … The machines, although they were working perfectly and so on, things had changed and computerization was starting to work its way in and the machines were expensive. Well it is all economics; it is nothing else.[22]

The redevelopment of the yard began in 1995, even before the change of ownership and direction within the firm. The company built thirty-eight new townhouses at Portside Court in the south yard, taking up over half of the original property.[23] The fitters' workshop and the joinery shop were demolished to make way for the new estate. The company's second venture in residential development had the advantage of better timing and was much more lucrative than its initial project at Kingston over a decade before. The new development, which appealed mainly to first-time buyers, exploited a niche in the market as very few houses had been built in the East Wall area over the previous decade.[24] The project was a significant commercial success, reflecting the early stages of a property boom, which would persist well into the first decade of the new millennium. The townhouses were publicly advertised for sale through the Ross McParland agency on 12 October 1995 and the entire estate was sold out within an hour of the launch.[25] Indeed Standish Collen,

---

22. Interview with Jimmy Small, 6 July 2009.

23. Collen Construction Ltd, Accounts, *Directors' Report for the year ended 31 March 1995*, p.3.

24. *Irish Times*, 'Fast sales at East Wall Road are reported', 19 October 1995.

25. *Irish Times*, 'East Wall Houses priced from £44,950', 12 October 1995; *Irish Times*, 'Notice – Portside Court, East Wall, Dublin 3', 19 October 1995; *Irish Times*, 'Fast sales at East Wall Road are reported', 19 October 1995.

who was rarely an advocate of residential development, was concerned that the townhouses had been sold off too quickly at an inadequate return. But the rest of the board did not share the misgivings of their most senior member – the directors of Collen Construction maintained the firm's customary reticence about giving profit figures for specific projects in their report for 1995–6, but noted that 'our townhouse development at East Wall has proved very successful.'[26] The project at Portside Court marked the beginning of a more sustained interest in developing houses than before, but on the company's own terms and without taking significant risks.

Following the change of ownership in 1996, the board embarked on a more ambitious programme of redevelopment. Neil Collen aimed to provide a modern headquarters for the company and was also interested in the potential for commercial development on the site, aiming to rent office space to commercial clients. The board of Collen Construction agreed on 28 August 1996 to seek planning permission for a new three-storey building in the north yard.[27] The proposed development received planning permission without difficulty, although initial negotiations to lease half of the new building to a commercial client were not successful.[28] The company itself took on the construction of the new development as a design and build project, negotiating temporary accommodation for its employees with Wiggins Teape, the paper merchants who owned the adjoining premises. The offices and the remaining sheds in the north yard were demolished, with the employees moving into the Wiggins Teape complex for several months. The new building, known as River House, was completed in December 1997.[29] River House opened for business from 2 February 1998, complete with the sculpture of two of the original Collen brothers over the doorway. Collen Construction occupied the top floor of the new building, while the other two floors were largely reserved for leasing to commercial clients.[30] SimulTrans, an international computer software company, was the first enterprise to take up office space in the new building, relocating its headquarters from the Harcourt Centre to River House in August 1998.[31]

---

26. Collen Construction, Accounts, *Directors' Report for the year ended 31 March 1995*, p.3.

27. Collen Construction, Minutes, Meeting of Directors, 28 August 1996, p.2.

28. Minutes, Meeting of Directors, 19 February 1997, p.2.

29. Minutes, Meeting of Directors, 14 May 1997, p.2; Minutes, Meeting of Directors, 10 December 1997, p.1.

30. Minutes, Meeting of Directors, 21 January 1998, p.1; Minutes, 15 July 1998, p.1.

31. Minutes, Meeting of Directors, 19 August 1998, p.3; *Irish Times*, 'New headquarters for software company', 2 September 1998.

Esat/British Telecom also leased accommodation on the ground floor towards the end of the same year. The redevelopment paid immediate dividends for the company, providing a valuable revenue stream from its commercial clients and highlighting the advantages of commercial development on its own property.

River House, East Wall Road. Courtesy of Collen Group.

The company evolved a commercial strategy in the late 1990s, which placed a high value on generating work for the different branches of the firm through its own development activity. The board of Collen Construction agreed on 21 January 1998 that 'the margins on tender work are still very competitive and it was considered desirable to produce our own work in preference to competitive tenders'.[32] Following the restructuring of the firm, Collen Group sought to obtain land not only to generate development profits but also to procure work for the construction and design companies. Neil Collen set an objective of generating a substantial element of the company's turnover from within its own resources:

32. Collen Construction, Minutes, Meeting of Directors, 21 January 1998, p.2.

We were trying to create work for ourselves as well as trying to find competitive work … there's obviously only a certain level of competitive work that you're actually able to get and there are resources within the group that we are able to go out and buy sites and develop them ourselves in a relatively conservative way, that's what we try to do.[33]

This approach was intended to allow the firm to initiate design and build projects, undertaken on a collaborative basis by the construction and project management companies. The redevelopment of the East Wall property, including the building of River House, fitted neatly into this strategy, even if it occurred before the board had formalized such an approach. So too did the much more ambitious development of the Wiggins Teape site between 2001 and 2007. The company undertook the redevelopment of the neighbouring site, which they purchased from Wiggins Teape in 1999.[34] The demolition of the original Wiggins Teape building by Collen in June 2001 aroused considerable controversy, involving protests from representatives of An Táisce, journalists and local residents in East Wall.[35] While Wiggins Teape was never a listed building, the opponents of the decision argued that it should have been listed and were critical of the local authority for failing to protect it.[36] The company, however, argued that its actions were not only legal but legitimate, on the basis that Wiggins Teape could not be left indefinitely as an empty building, it was not in fact a neo-classical building as it dated from 1930, and part of the entrance portico was being retained.[37] Following appeals by the parties to An Bord Pleanála, the proposed office development received planning permission in 2002.[38]

---

33. Interview with Neil Collen, 28 April 2009.

34. Collen Construction, Minutes, Meeting of Directors, 14 January 1999, p.2; Minutes, Meeting of Directors, 23 February 1999, p.2.

35. *Irish Times*, 'Planning: Decisions to Grant Permission', 4 May 2000; *Irish Times*, 'Collen plans large four-block East Wall development', 10 May 2000; *Irish Times*, 'Artist fights to save factory from demolition', 16 November 2000; *Irish Times*, 'Neo-classical factory demolished in spite of An Bord Pleanála finding', 18 June 2001; An Bord Pleanála, Decision on Planning Appeal, 119524: (0266/00), 12 June 2001.

36. *Irish Times*, 'Architectural heritage is dying on the vine of ignorance, ineptitude and sheer cowardice', 28 June 2001.

37. Interview with Leo Crehan, 11 February 2001; *Irish Times*, 'Artist fights to save factory from demolition', 16 November 2000.

38. *Irish Times*, 'Collen Group gets go-ahead for offices on Wiggins Teape site', 23 January 2002; Collen Construction, Meeting of Directors, 25 April 2002; An Bord Pleanála, Decision on Planning Appeal, 129627: (0047/02), 26 June 2002.

The company's application to build apartments at the rear of the same site was approved at the end of the following year.[39] Collen Group first initiated a large-scale residential development on the Wiggins Teape site, providing two blocks of 157 apartments which were completed between 2004 and 2006.[40] Then a new five-storey office building was constructed on the front end of the site between 2005 and 2007: the new One Gateway building served a dual function as the headquarters for Collen Group and the largest commercial development so far undertaken by the firm. While Collen Group itself occupied the top floor, most of the new building was reserved for office space, to be rented to commercial clients.[41] Indeed it was in commercial development that the company initiated its most striking new departure during the boom. The redevelopment of the company's East Wall property and the building of One Gateway marked a major investment in the development of office space, which went considerably further than previous initiatives by the firm in this area.

The Gateway buildings, East Wall Road. Courtesy of Collen Group.

39. An Bord Pleanála, Decision on Planning Appeal, 205358: (3209/03), 22 December 2003; Collen Group, Minutes, Meeting of Directors, 11 December 2003.

40. Collen Group, Minutes, Meeting of Directors, 11 December 2003, p.2; Collen Group, Minutes, Meeting of Directors, 10 June 2004, p.2.

41. Interview with Paddy Wall, 14 January 2010.

Yet the new emphasis on development by the group company did not mean that its directors had thrown caution to the winds. Martin Glynn believed that cautious business practice remained an enduring element of the company's approach:

> There would always be that conservatism, you know any risk we took was well risk researched and evaluated and it might appear as a risk but I don't think it would be. I would say the firm would always be a conservative firm that they certainly wouldn't go on a project that might lead them into trouble.[42]

Several of the residential projects initiated by the company were built on its own land, which did much to control costs and reduce the financial risks associated with buying and developing land. The success of Portside Court illustrated the advantages of development on the company's property. The company also bought land for development elsewhere, either in conjunction with partners or independently. Collen Construction built a new estate of townhouses at Willbrook Road, Rathfarnham, in 1999, on a site developed by Jindsberg Construction, a subsidiary company part-owned by Neil Collen. The estate consisted of thirty-five two- or three-bedroom houses set around a landscaped central courtyard with small open-plan gardens.[43] Collen Group itself developed a new housing estate at Malahide four years later. The company secured planning permission in June 2003 for a new development of fifty-three houses, duplex and apartment units on a site at Lissadel Wood (off the Swords road in Malahide), which Collen had purchased for development.[44] The new estate was completed by Collen Construction in 2005, although the first two phases of the development were launched in the autumn of the previous year.[45] Residential development, however, accounted for a much smaller element of the company's turnover than commercial development or tendered projects.

While the company expanded its involvement in development during the boom, Collen had little in common with the first division of major property developers. The scale of its activity in the property market, particularly residential development, was much less extensive than that of major builders. Moreover, commercial or residential development was not simply an end in itself, but was intended to benefit the core business of the company. Collen

---

42. Interview with Martin Glynn, 16 June 2009.

43. *Irish Times*, '£190,000 townhouses in Rathfarnham', 25 November 1999.

44. Collen Group, Minutes, Meeting of Directors, 19 June 2003.

45. Collen Construction, Minutes, Meeting of Directors, 18 May 2004, p.1; Minutes, 9 December 2004, p.1.

Group had different objectives in acquiring land from conventional property developers, aiming at least in part to generate turnover for the construction and design companies. The company had neither the inclination nor the opportunity to engage in the large-scale purchase and development of land for its own sake. The firm might not have shared in the most lucrative gains of the property boom, but managed to avoid becoming dependent on property development.

The company in practice developed a diversified portfolio that included both tendered and negotiated contracts, as well as projects generated through the group's development activity. Collen Construction, in particular, secured a wide variety of contracts during the Celtic Tiger era, undertaking projects for both public authorities and private clients. The building company was deeply involved in delivering administrative, educational and cultural projects for several government departments, the OPW and other public institutions. The company returned to its traditional haunts in the Curragh at the beginning of the twenty-first century, winning various contracts with the Department of Defence. Collen provided a new mess for non-commissioned officers and refurbished the Combat Support College at the camp between 2001 and 2002.[46] The construction of schools formed a significant element of Collen Construction's turnover in the first decade of the twenty-first century. The company completed a series of new school buildings, including a new community school at Kilcoole, built for the Co. Wicklow VEC, a national school in Celbridge for the Department of Education, and new primary and secondary schools in Lucan.[47] Collen also acquired business in the university sector, building student residences at Roebuck Hall in UCD and refurbishing the President's Lodge in 2004.[48] The firm had built schools intermittently in previous decades, but established a strong association with public educational projects during the Celtic Tiger era.

The company played a significant part in the refurbishment or extension of historic public buildings during this period. Collen Construction refurbished an existing stable block at Collins Barracks to provide laboratory space, having won the tendering competition with the OPW in 2000. The company later added a new extension to the barracks to accommodate a military history

---

46. Collen Construction, Minutes, Meeting of Directors, 18 January 2001, p.1; Minutes, 29 August 2002, p.1.

47. Collen Construction, Minutes, Meeting of Directors, 5 December 2001, p.2; Minutes, 14 June 2002, p.1; Minutes, 18 September 2003, p.1.

48. Collen Construction, Minutes, Meeting of Directors, 25 March 2004, p.1.

exhibition, complete with a display of military aircraft, which was intended to serve a similar purpose to the Imperial War Museum in London.[49]

The building company undertook various refurbishment or conservation projects on public buildings in Dublin, including the Custom House, where an internal renovation was required to establish a visitors' centre.[50] Collen Construction also took charge of the refurbishment and extension of the Hugh Lane municipal gallery at Charlemont House, Parnell Square, working in conjunction with Gilroy McMahon architects.[51] The extension of the gallery between 2004 and 2006 was a particularly significant project, which was valued at £11 million. It marked the high point of the company's involvement in the renovation of prestigious civic buildings.

Roebuck Hall, UCD. Courtesy of Collen Construction.

---

49. Collen Construction, Minutes, Meeting of Directors, 18 January 2000, p.2; Minutes, 20 September 2001, p.2; Minutes, 12 January 2006, p.1; Interview with Leo Crehan, 29 June 2009.

50. Interview with Leo Crehan, 29 June 2009; Collen Construction, Minutes, Meeting of Directors, 13 October 2005, p.1.

51. Collen Construction, Minutes, Meeting of Directors, 18 May 2004, p.1.

The extension to the Hugh Lane Gallery. Courtesy of Collen Construction.

The provision of social and affordable housing emerged as a key strand of activity for the construction company during the later stages of the economic boom. The firm had not built residential housing for public or non-profit institutions on a significant scale since the late 1940s, but revived a long dormant tradition early in the new century. Collen Construction enjoyed considerable success in tendering for social housing projects with local authorities. The company undertook several housing schemes for Dún Laoghaire and Rathdown county council. The most substantial of these initiatives was a social and affordable residential development in Goatstown, which was valued at €17 million. The project, incorporating ninety-three units, including houses and apartments, was completed early in 2010.[52] The building company undertook several other public housing projects in Dún Laoghaire, including a scheme involving the building of local authority houses on Laurel Avenue, Ballybrack and a smaller social and affordable housing project at Meadowlands, incorporating a modern community centre: the construction of the housing for the latter project is still in progress.[53] While tendering for public housing contracts undoubtedly generated the greatest rewards, the company also secured business

---

52. Interview with Martin Glynn, 16 June 2009; Collen Construction, Minutes, Meeting of Directors, 13 December 2007, p.1.

53. Interview with Leo Crehan, 11 February 2010.

from voluntary housing associations for the first time since the 1940s. Collen Construction undertook two social housing projects for *Respond*, a leading non-profit housing association – the company built houses, apartments and a community centre in the Tolka Valley, while also completing a new *Respond* housing scheme on the East Road not far from the company's headquarters in East Wall.[54] Social housing became a much more substantial and valuable element of the building company's turnover than private residential housing in the first decade of the twenty-first century.

Collen Construction also secured a considerable volume of business in the private sector, much of it achieved on the basis of negotiation rather than tendering. The building company acquired a great deal of work from Green Property, which was developing the Fonthill industrial estate in Clondalkin: Collen Construction negotiated terms directly with Green Property and built a substantial share of the industrial units on the estate between 1999 and 2005.[55] Motor Services Ltd emerged as another key client, in this instance on the

MSL Service Centre, Pottery Road. Courtesy of Collen Construction.

54. Collen Construction, Minutes, Meeting of Directors, 25 April 2002, p.1; Interview with Leo Crehan, 29 June 2009.

55. Collen Construction, Minutes, Meeting of Directors, 20 May 1999, p.2; Minutes, 22 May 2003, p.1; Minutes, 12 October 2004, p.1.

The Park, Carrickmines. Courtesy of Collen Project Management.

basis of the company's success in winning tenders. The construction company built a considerable number of motor-show rooms and repair centres for MSL, including new facilities in Ballsbridge, Deansgrange, and on Pottery Road in Dún Laoghaire.[56] The company's activity in the private sector showed a roughly even split between tendered projects and negotiated work. The company's ability to secure a reasonable level of negotiated work was an important asset at a time when tendering operated on extremely competitive margins that tended to favour major developers or building conglomerates.

Meanwhile, CPM developed a flourishing client base of its own. The design company undertook a number of projects for Park Developments. CPM designed a new commercial development at Carrickmines between 2002 and 2004, which was located beside one of the junctions of the new M50. The project (known as The Park, Carrickmines) incorporated retail units, warehouses and office space.[57] The design company also took charge of the planning of a new business park for the same client on a hundred-acre site in Mitchelstown and Ballycoolin.[58] CPM maintained its business association with SIAC, which had been forged shortly before the design company's emergence as an autonomous entity, taking responsibility for the design of industrial and

56. Collen Construction, Minutes, Meeting of Directors, 18 January 2000, p.2; Minutes, 18 January 2001, p.1.

57. CPM, Minutes, Meeting of Directors, 5 February 2002, p.1; Minutes, 24 January 2004, p.2; Minutes, 29 September 2004, p.1.

58. CPM, Minutes, Meeting of Directors, 15 February 2001, p.2.

commercial units (developed and built by SIAC) in Baldonnel Business Park.[59] Residential design also formed a significant part of the company's activity. CPM designed several phases of a new housing development for Dolminack Ltd at Meakstown, near Finglas, between 2003 and 2005: the project consisted of over six hundred units, including a mixture of apartments, duplex housing and semi-detached townhouses.[60] The restructuring of the group undoubtedly facilitated the design company in winning projects that it might not have secured otherwise, opening up opportunities which would not have been available if Collen was committed primarily to a design and build approach.

While major building contractors and developers undoubtedly formed a key element of its business, CPM also sought clients outside the traditional environs of the construction industry. The company designed a new credit union at Dundrum in 2004.[61] The design company also found fertile

Dundrum Credit Union. Courtesy of Collen Project Management.

59. Collen Construction, Minutes, Meeting of Directors, 14 May 1997, p.2; CPM, Minutes, Meeting of Directors, 31 July 2002, p.1.

60. *Irish Times*, 'Fingal: Decision to Grant', 17 July 2003; *Irish Times*, 'Fingal: Applications', 5 February 2004; Fingal County Council, Minutes, 25 September 2008; CPM, Minutes, Meeting of Directors, 17 July 2003, p.1; Minutes, 25 September 2003, p.1; Minutes, 25 May 2004, p.1.

61. CPM, Minutes, Meeting of Directors, 25 May 2004, p.1.

RDS Sports Grounds, Ballsbridge. Courtesy of Collen Project Management.

opportunities with one of Collen's most established clients, the RDS. CPM designed a new extension to Shelbourne Hall for the Society early in the twenty-first century, providing greater space for exhibitions at a venue originally built by Collen Brothers.[62] More recently, the company oversaw the design of new stands for the rugby grounds at the RDS. It was striking that CPM relied exclusively on private clients, in contrast to the building company's extensive involvement in public service projects. This situation reflected the recent origins of the autonomous design company and the trend for public contracts to be won by larger architectural or design practices. Despite the very different client base and business profile established by the construction and design companies, each shared some key characteristics, not least the capacity to retain the loyalty of key clients. The continuing ability of both companies to attract repeat business underlined that in an era where traditional loyalties often counted for little, Collen's ability to forge a lasting association with many of its clients remained intact.

The two major wings of the firm operated independently, sometimes co-operating to undertake design and build contracts but usually acquiring and completing their own projects. Collen Construction drew about 70 per cent of its turnover from building projects for external clients in the decade following the restructuring – design and build projects accounted for less than

62. Interview with Chris Lyons, 6 July 2009.

a third of its activity.[63] This pattern was even more marked for CPM. Chris Lyons noted that the vast majority of the design company's work during the economic boom was to be found among external clients: 'over 80 per cent of design work would have been for outside clients'.[64] Much of the design and build work occurred in the context of developments initiated by Collen Group itself. The redevelopment of the East Wall property was the most striking example of collaboration between the two companies. Portside Court and Collen Construction's new headquarters at River House were both design and build projects. More recently, the One Gateway building and the neighbouring apartment blocks also formed part of a collaborative effort involving the two major divisions of the firm. The two companies collaborated too on several projects not directly associated with the group's own development activity. Perhaps the most notable was a new national school at Rowlestown in north Dublin: the firm took charge of the design and construction of a modern school building, along with a sports hall and an all-weather pitch, which was essentially completed by August 2008.[65] But aside from development ventures initiated by the group itself, design and build projects were increasingly the exception rather than the norm during the Celtic Tiger.

Rowlestown National School, north Dublin. Courtesy of Collen Construction.

63. Interview with Leo Crehan, 11 February 2010.

64. Interview with Chris Lyons, 6 July 2009.

65. Collen Construction, Minutes, Meeting of Directors, 16 August 2007, p.2; Minutes, 23 October 2008, p.2.

The readiness of the board under Neil Collen's direction to engage in new business ventures and an ambitious redevelopment programme did not mean a fundamental break with the past. The directors and managers were more open to calculated risk-taking, but they did not abandon the pragmatic caution or low-key modus operandi displayed by their predecessors. The company did not seek to compete for tenders on razor-thin margins, nor did its managers show any inclination to gamble on the basis of the spiralling value of development land. While the company was undoubtedly more willing to take commercial risks during the economic boom and engaged in a wider selection of development activity than ever before, the Celtic Tiger did not cause Collen to jettison its distinctive values. The firm retained its essential character as a family business, although responsibility and authority were much more widely diffused than previously. The company also maintained a diversified client base, not relying solely on the uncertain vagaries of the property market. The diversified character of Collen's business proved an invaluable asset when the international economy entered a tailspin in 2008. The directors of Collen embraced the opportunities offered by the Celtic Tiger, but did not succumb to the temptation to gamble the future of their business on its indefinite continuation.

Postscript

Collen Brothers began as a small building firm in north Armagh. With the establishment of a builders' yard in Clanwilliam Place, the company became a firm fixture of the commercial scene in Dublin and had, within a generation, extended its tentacles to most regions of the country. The violent collapse of British rule presented the first real crisis of the firm's existence. Collen displayed a high level of resilience and flexibility in adapting to the partition of the island and in surviving as an all-Ireland business for a quarter of a century. The international economic depression did not prevent Collen Brothers from maintaining a viable business on both sides of the border, depending on a combination of state contracts and repeat business from traditional private clients in the difficult economic conditions of the 1930s. Yet the wartime experience of the two branches of the firm diverged significantly, with one fulfilling the exceptional demands of the Allied military building programme in Northern Ireland, and the other adapting to the more limited requirements of a neutral, protectionist state. The division of the firm in 1949 was influenced by a generational transition, but also marked a realistic recognition that the two branches of the business had to be allowed to develop in their own way.

Religious or political allegiance did not influence how or where the company did business at any stage, although John Collen, the driving force behind the foundation of the company and its dominant figure in the late nineteenth century, was a dedicated Unionist. The firm in Northern Ireland benefited extensively from public contracts under the Stormont government, but was equally successful in securing business from the administration established by the British government under direct rule. The company weathered the Troubles successfully, at least in part because its directors assiduously avoided identification with any political or cultural agenda. Ultimately it was not the political environment but the initiatives taken by the proprietors in response to unfavourable economic circumstances that led to the reshaping of the business.

Collen's status as a Protestant family business south of the border undoubtedly influenced its development and was more of an asset than a liability in the independent Irish state. Collen benefited from an influential network of Protestant commercial enterprises in Dublin from the late 1880s until well into the twentieth century, although it was never completely dependent on such connections. Moreover, a striking element of the company's development, especially since the 1960s, is the extent to which it broke out from this traditional milieu and developed an impressively wide client base. The scope and diversity of Collen's activity, which first became apparent during the post-war era, remained an important strength of the business in the twenty-first century.

The two companies that emerged from the traditional Collen family enterprise evolved in very different directions, although their response to external economic conditions showed common ground. The two Collen companies faced comparable economic pressures between the late 1970s and early 1990s: both responded with extensive programmes of rationalization and initiated radical changes in the character of their business – such changes ultimately sharpened the distinction between the two firms, as Collen Brothers in Northern Ireland concentrated on its core business in quarrying, while Collen Construction (and later Collen Group) focused on building and property development. Interestingly, major civil engineering projects were gradually phased out due to competitive pressures on both sides of the border.

It is too early to attempt any historical evaluation of the long-term consequences of the economic crash in 2008 for Irish society in general or Collen in particular. The immediate impact of the crisis on the Irish economy was particularly severe because it was combined with the inevitable collapse of the domestic property bubble, inflated largely by flawed policy decisions. The collapse of demand in the construction industry was one of the most obvious symptoms of the recession. Collen's diverse client base and realistic business model, however, has served it well in the past and is likely to do so again.

Collen Brothers was a 'classic' family business of the kind identified by Andrea Colli, in which ownership and control were firmly intertwined, with family members being involved in both strategic and day-to-day decision-making.[1] Indeed, the company at the outset was not only firmly controlled by family directors but also relied heavily on family members to manage the firm on an operational basis. This sustained involvement by family members was a traditional feature of the business, with Joseph Collen's personal supervision of the building of Killarney House during the 1870s being only the most obvious example. It was a trend that was still apparent early in the twentieth century, with Harky and Jack Collen taking a leading role in managing the Portrane project and Harky's direction of the firm's operations at the Curragh. While greater delegation in overseeing specific projects has become the norm, the family has retained a central position in managing the business on both sides of the border. This pattern was particularly marked in the original firm following the division of the business in 1949. Collen Brothers in Portadown retained a managerial structure characteristic of a traditional family business, in which the proprietors not only gave overall direction to the business but also oversaw its day-to-day management. Collen Brothers Dublin adopted a different approach, based on selective delegation to a small number of key figures: the company developed a recognizable upper managerial cohort within its head office in the late 1960s. Yet the predominant role of Standish and Lyal Collen was universally acknowledged and the scope for independent action by professional managers was extremely limited. Standish in particular determined the extent of delegation, maintaining a tight control over the management of the business. Although Collen Group recently adopted an institutional structure that gives considerable autonomy to subsidiary companies and key managers, the reorganization also preserved the pivotal position of the family proprietors in management and strategic decision-making, a defining feature of the firm both north and south.

The issue of leadership succession presents particular challenges for any family business.[2] Collen evolved a distinctive process of internal succession, which was characterised by long-term planning and careful preparation for the delegation of authority to younger members of the family. The arrangements

---

1. Colli, *History of Family Business*, p.9.

2. *Ibid.* pp.66–9.

that ultimately prevailed were generally determined on the basis of consensus among family directors active within the business. Disagreement over ownership and succession sometimes caused considerable tension: this was apparent shortly after the partition of the island in the 1920s and again in a more severe form during the 1940s, when a dispute over the ownership of shares triggered a sharp divergence among the directors. But such severe disagreements were rare and family members were usually able to settle their differences. Moreover, any internal disputes were ultimately resolved without threatening the survival of the firm itself. It is striking that the division of the business in 1949 was accomplished by consensus between the two branches of the family. The key business decisions in both companies for the rest of the twentieth century were also usually achieved by agreement between the participants. The Collen family, before and after the division of the firm, displayed a strong sense of solidarity and a notable commitment to the preservation of the business, which was sufficiently compelling to override individual rivalries or differences.

The family directors, past and present, have been keenly aware that the firm or its successor companies would never have flourished without the commitment and dedication of their workers. Joe Collen was only one of many directors who paid tribute to the unsung labours of construction workers, who performed a wide variety of jobs in frequently arduous conditions. A strong element of benign paternalism was evident in the approach of the proprietors towards their employees for much of the twentieth century, as the family directors valued the loyalty of their staff and in turn recognised reciprocal obligations for their welfare. While this paternalistic approach has ceased to be significant with the passage of time, the companies on both sides of the border have maintained a strong sense of internal cohesion based on a culture of community and mutual support between directors, managers and employees.

From the beginning, Collen developed as a medium-sized company, alternating between peaks of rapid expansion and troughs of severe contraction, but never becoming a large-scale corporation. Economic conditions played their part in determining the development of the company, but the aspirations and preferences of the proprietors were arguably crucial in deciding the size and reach of the firm. Collen Brothers Dublin undoubtedly had the potential to develop into a much larger enterprise during the 1970s. When the company greatly extended both the range of its activity and the value of its turnover during this period, it acquired the capital to underpin further expansion. Moreover, its collaboration with Christiani & Nielsen at Aughinish Island could have served as a forerunner for other collaborative ventures in Britain or further afield. But the company, largely due to the influence of Standish Collen, did not take the opportunity to extend its operations on an international scale. This was a deliberate decision that reflected not only Standish's legendary caution, but also his reluctance to sanction ventures that would not be under the direct control of the family. Yet the company's development was not simply a matter of Standish's personal preferences: the proprietors throughout the company's history were deeply committed to a business model in which the interests of the family and company were inextricably intertwined. Collen never showed any real indication of evolving into a large-scale managerial corporation of the kind identified by Chandler, characterized by a series of distinct operating units and managed by a hierarchy of salaried managers.[3]

Yet if Collen's character as a family business militated against relentless expansion, it also promoted a fundamental stability within the firm. The distinctive traditions of the Collen family found expression in the core values of the proprietors, which remained remarkably consistent over time – pragmatism, scepticism towards excessive risk-taking, a marked preference for a low-key method of operation and a determination to establish a long-term relationship with clients. The family directors valued the profits from the business, but gave precedence to the survival of the firm as a stable and commercially viable institution rather than the maximization of immediate financial return. But stability did not mean stagnation. The family directors have always displayed the ability to adapt to changing political or economic circumstances (not least following the partition of the island in 1922), jettisoning traditional company structures, phasing out previously valuable operations and even – at least in Dublin – dropping the established name of Collen Brothers itself. Time and time again, Collen showed an ability to renew itself: in Portadown in the late 1970s, in Dublin during the following decade, and again with the creation of the new group structure in the 1990s. The history of Collen offers persuasive evidence that a versatile and resilient family business can establish an enduring niche in a modern industrialized economy.

---

3. Chandler, 'The United States: Seedbed of Managerial Capitalism', in Chandler and Daems (ed), *Managerial Hierarchies: Comparative Perspectives on the Rise of the Modern Industrial Enterprise*, pp.13–14.

Bibliography

STATE ARCHIVES AND RECORDS
Companies Registration Office (CRO)
Annual returns submitted by Collen Brothers (Dublin) Ltd, 1963–84
Resolution and other papers relating to the voluntary liquidation of Collen Brothers, 1985–1990
Accounts for Collen Construction, 1989–2008
Accounts for Collen Group, 1997–2008

IRISH ARCHITECTURAL ARCHIVE (IAA)
Accession 2009/100, Album of press cuttings relating to Vincent Kelly
Patterson, Kempster and Shortall Collection: account books and letter books
Scott, Tallon and Walker Drawings Collection, vol.1: Scott and Good records

NATIONAL ARCHIVES OF IRELAND (NAI)
Census of Ireland, 1911

NORTHERN IRELAND COMPANIES OFFICE
Certificate of incorporation for Collen Brothers (Quarries) Ltd

PAMPHLETS
Dublin Corporation and County Council, *The Greater Dublin Drainage Scheme* (Dublin, 1986)

PRIVATE COLLECTIONS
Collen Papers, Hanover Street, Portadown
Collen Papers, River House Archive, East Wall Road
Lyal Collen Papers
Joseph Collen records
Minutes of Collen Brothers, Dublin Ltd, 1950–84

Minutes of Collen Brothers Quarries Ltd, 1978–2007
Minutes of Collen Construction Ltd, 1984–2008
Minutes of Collen Group Ltd, 1997–2008
Minutes of Collen Project Management Ltd, 1999–2008

JOURNALS
The Architects' Journal
Irish Geography
The Irish Builder and Engineering Record, 1869–1965
The First Irish Roads Congress, Records of Proceedings (1910)
Review: Journal of the Craigavon Historical Society

NEWSPAPERS AND MAGAZINES
Business and Finance
The Irish Independent
The Irish Press
The Irish Times
The Sunday Independent
The Leinster Leader
The Portadown Times

SECONDARY WORKS
Bence-Jones, Mark, *A Guide to Irish Country Houses* (London, 1996)
Berry, H.F., *A History of the Royal Dublin Society* (London, 1915)
Bond, B.L., *Case History: Main Lift Pumping Station at Ringsend, Dublin* (unpublished presentation, March 2000)
Bond, B.L., *Upheaval Pressure and Base Failure*, in E.T. Hanrahan, T.L.L. Orr and T.F. Widdis, *Proceedings of the Ninth European Conference on Soil Mechanics and Foundation Engineering*, 31 August–3 September 1987
Bradshaw, Thomas, *Bradshaw's General Directory of Newry, Armagh and the towns adjoining for 1820* (Newry, 1820)
Chandler, Alfred, 'The United States: Seedbed of Managerial Capitalism', in Alfred Chandler and Herman Daems (eds), *Managerial Hierarchies: Comparative Perspectives on the Rise of the Modern Industrial Enterprise* (Harvard, 1980), pp.9–40
Clarke, Desmond and Meenan, James, *The Royal Dublin Society 1731–1981* (Dublin, 1981)
Coleman, Marie, *The Irish Sweep: A History of the Irish Hospitals Sweepstake 1930–87* (Dublin, 2009)
Colli, Andrea, *The History of Family Business: New Studies in Economic and Social History* (Cambridge, 2003)
Colli, Andrea and Rose, Mary, *Families and Firms: The Culture and Evolution of Family Firms in Britain and Italy in the Nineteenth and Twentieth Centuries, Scandinavian Economic History Review*, vol.47. no.1 (1999), pp.24–47
Collen, Lyal, 'Building' in *Careers in Ireland* (Dublin, 1958), pp.16–19

Costello, Con, *A Most Delightful Station: The British Army on the Curragh of Kildare, Ireland, 1855–1922* (Cork, 1996)

Cox, Ronald, *Civil Engineering at Trinity: A Record of Growth and Achievement* (Dublin, 2009)

De Vere White, Terence, *The Story of the Royal Dublin Society* (Tralee, 1955)

Ferriter, Diarmuid, *The Transformation of Ireland 1900–2000* (London, 2004)

Flanagan, P.J., *The Cavan and Leitrim Railway* (Newton Abbot, Devon, 1966)

Gilligan, H.A., *A History of the Port of Dublin* (Dublin, 1988)

Hoffman, Ronnie, 'It took 200 years to build', *Business and Finance*, vol.17, no.33, 30 April 1981

Lee, Joseph, *Ireland 1912–85: Politics and Society* (Cambridge, 1989)

Little, A.L., Bond, B.L., and Marshall, W.J., *Groundwater Control in the construction of a Dublin pumping station*, in E.T. Hanrahan, T.L.L. Orr and T.F. Widdis (eds), *Proceedings of the Ninth European Conference on Soil Mechanics and Foundation Engineering: Groundwater Effects in Geotechnical Engineering*, 31 August–3 September 1987, pp.183–8

Lyons, F.S.L., *Ireland since the Famine* (London, 1972)

McManus, Ruth, 'The "Building Parson" – The role of Reverend David Hall in the solution of Ireland's early twentieth century housing problems', *Irish Geography*, vol. 32(2), 1999, pp.87–98

Murphy, Gary, 'From economic nationalism to European Union', in Girvin, Brian and Murphy, Gary, *The Lemass Era: Politics and Society in the Ireland of Seán Lemass*, pp.28–48 (Dublin, 2005), pp.28–48

Murphy, Gary, *In Search of the Promised Land: The Politics of Post-War Ireland* (Cork, 2009)

O'Brien, Justin, *The Arms Trial* (Dublin, 1990)

O'Sullivan, P.M., 'Recent development works in Dublin Port', *Proceedings, The Institution of Civil Engineers, Supplement*, 1970 (vii), pp.181–2

Rothery, Seán, *Ireland and the new architecture 1900–1940* (Dublin, 1991)

Rowan, Ann Martha, *A Dictionary of Irish Architects 1720–1940* (IAA, Dublin, 2005)

INTERVIEWS

Dr Brian Bond, 30 April 2009; 11 January 2010; 5 February 2010; 30 January 2010

Louise Coffey, 23 June 2009

David Collen, 10 June 2009

Neil Collen, 28 April 2009

Niall Collen, 14 May 2009; 5 February 2010; 16 March 2010

Joe Collen, 14 May 2009

Marcus Collie, 23 June 2009

Dr Ronald Cox, 30 July 2009

Leo Crehan, 29 June 2009; 11 February 2010

Martin Glynn, 16 June 2009

Des Lynch, 9 June 2009

Chris Lyons, 6 July 2009

Rita McMillan, 13 July 2009
Jerry O'Leary, 22 July 2009
Frank O'Sullivan, 1 July 2009
John Ruane, 16 July 2009
Pat Sides, 21 July 2009
Jimmy Small, 6 July 2009
Paddy Wall, 28 April 2009; 28 July 2009; 14 January 2010

Index

Page numbers in *italics* indicate pages with photographs.